Preface

This book contains the solutions of all the exercises of my book: Principles of Tensor Calculus. These solutions are sufficiently simplified and detailed for the benefit of readers of all levels particularly those at introductory levels.
Taha Sochi
London, September 2018

Contents

Preface — 1

Table of Contents — 2

Nomenclature — 3

1 Preliminaries — 6

2 Spaces, Coordinate Systems and Transformations — 15

3 Tensors — 34

4 Special Tensors — 53

5 Tensor Differentiation — 83

6 Differential Operations — 113

7 Tensors in Application — 130

Index — 153

Nomenclature

In the following list, we define the common symbols, notations and abbreviations which are used in the book as a quick reference for the reader.

∇	nabla differential operator
∇_i and ∇^i	covariant and contravariant differential operators
∇f	gradient of scalar f
$\nabla \cdot \mathbf{A}$	divergence of tensor $\mathbf{A}$
$\nabla \times \mathbf{A}$	curl of tensor $\mathbf{A}$
$\nabla^2, \partial_{ii}, \nabla_{ii}$	Laplacian operator
$\nabla \mathbf{v}, \partial_i v_j$	velocity gradient tensor
, (subscript)	partial derivative with respect to following index(es)
; (subscript)	covariant derivative with respect to following index(es)
hat (e.g. $\hat{A}_i, \hat{\mathbf{E}}_i$)	physical representation or normalized vector
bar (e.g. $\bar{u}^i, \bar{A}_i$)	transformed quantity
$\circ$	inner or outer product operator
$\perp$	perpendicular to
1D, 2D, 3D, nD	one-, two-, three-, n-dimensional
$\delta/\delta t$	absolute derivative operator with respect to t
∂_i and ∇_i	partial derivative operator with respect to i^{th} variable
$\partial_{;i}$	covariant derivative operator with respect to i^{th} variable
$[ij,k]$	Christoffel symbol of 1st kind
A	area
$\mathbf{B}, B_{ij}$	Finger strain tensor
$\mathbf{B}^{-1}, B_{ij}^{-1}$	Cauchy strain tensor
C	curve
C^n	of class n
$\mathbf{d}, d_i$	displacement vector
det	determinant of matrix
diag $[\cdots]$	diagonal matrix with embraced diagonal elements
$d\mathbf{r}$	differential of position vector
ds	length of infinitesimal element of curve
$d\sigma$	area of infinitesimal element of surface
$d\tau$	volume of infinitesimal element of space
$\mathbf{e}_i$	i^{th} vector of orthonormal vector set (usually Cartesian basis set)
$\mathbf{e}_r, \mathbf{e}_\theta, \mathbf{e}_\phi$	basis vectors of spherical coordinate system
$\mathbf{e}_{rr}, \mathbf{e}_{r\theta}, \cdots, \mathbf{e}_{\phi\phi}$	unit dyads of spherical coordinate system
$\mathbf{e}_\rho, \mathbf{e}_\phi, \mathbf{e}_z$	basis vectors of cylindrical coordinate system
$\mathbf{e}_{\rho\rho}, \mathbf{e}_{\rho\phi}, \cdots, \mathbf{e}_{zz}$	unit dyads of cylindrical coordinate system
$\mathbf{E}, E_{ij}$	first displacement gradient tensor
$\mathbf{E}_i, \mathbf{E}^i$	i^{th} covariant and contravariant basis vectors

$\underline{\mathbf{E}}_i$	i^{th} orthonormalized covariant basis vector
Eq./Eqs.	Equation/Equations
g	determinant of covariant metric tensor
$\mathbf{g}$	metric tensor
g_{ij}, g^{ij}, g_i^j	covariant, contravariant and mixed metric tensor or its components
$g_{11}, g_{12}, \cdots g_{nn}$	coefficients of covariant metric tensor
$g^{11}, g^{12}, \cdots g^{nn}$	coefficients of contravariant metric tensor
h_i	scale factor for i^{th} coordinate
iff	if and only if
J	Jacobian of transformation between two coordinate systems
$\mathbf{J}$	Jacobian matrix of transformation between two coordinate systems
$\mathbf{J}^{-1}$	inverse Jacobian matrix of transformation
L	length of curve
$\mathbf{n}, n_i$	normal vector to surface
P	point
$P(n, k)$	k-permutations of n objects
q^i	i^{th} coordinate of orthogonal coordinate system
$\mathbf{q}_i$	i^{th} unit basis vector of orthogonal coordinate system
$\mathbf{r}$	position vector
$\mathcal{R}$	Ricci curvature scalar
R_{ij}, R_j^i	Ricci curvature tensor of 1st and 2nd kind
R_{ijkl}, R^i_{jkl}	Riemann-Christoffel curvature tensor of 1st and 2nd kind
r, θ, ϕ	coordinates of spherical coordinate system
S	surface
$\mathbf{S}, S_{ij}$	rate of strain tensor
$\bar{\mathbf{S}}, \bar{S}_{ij}$	vorticity tensor
t	time
T (superscript)	transposition of matrix
$\mathbf{T}, T_i$	traction vector
tr	trace of matrix
u^i	i^{th} coordinate of general coordinate system
$\mathbf{v}, v_i$	velocity vector
V	volume
w	weight of relative tensor
x_i, x^i	i^{th} Cartesian coordinate
x'_i, x_i	i^{th} Cartesian coordinate of particle at past and present times
x, y, z	coordinates of 3D space (mainly Cartesian)
$\boldsymbol{\gamma}, \gamma_{ij}$	infinitesimal strain tensor
$\dot{\boldsymbol{\gamma}}$	rate of strain tensor
Γ_{ij}^k	Christoffel symbol of 2nd kind
$\boldsymbol{\delta}$	Kronecker delta tensor
$\delta_{ij}, \delta^{ij}, \delta_i^j$	covariant, contravariant and mixed ordinary Kronecker delta
$\delta_{kl}^{ij}, \delta_{lmn}^{ijk}, \delta_{j_1\ldots j_n}^{i_1\ldots i_n}$	generalized Kronecker delta in 2D, 3D and nD space

$\boldsymbol{\Delta}$, Δ_{ij}	second displacement gradient tensor
ϵ_{ij}, ϵ_{ijk}, $\epsilon_{i_1...i_n}$	covariant relative permutation tensor in 2D, 3D and nD space
ϵ^{ij}, ϵ^{ijk}, $\epsilon^{i_1...i_n}$	contravariant relative permutation tensor in 2D, 3D and nD space
$\underline{\epsilon}_{ij}$, $\underline{\epsilon}_{ijk}$, $\underline{\epsilon}_{i_1...i_n}$	covariant absolute permutation tensor in 2D, 3D and nD space
$\underline{\epsilon}^{ij}$, $\underline{\epsilon}^{ijk}$, $\underline{\epsilon}^{i_1...i_n}$	contravariant absolute permutation tensor in 2D, 3D and nD space
ρ, ϕ	coordinates of plane polar coordinate system
ρ, ϕ, z	coordinates of cylindrical coordinate system
$\boldsymbol{\sigma}$, σ_{ij}	stress tensor
$\boldsymbol{\omega}$	vorticity tensor
Ω	region of space

Chapter 1
Preliminaries

1. Differentiate between the symbols used to label scalars, vectors and tensors of rank > 1.
   Answer:[1]
   Scalars: non-indexed lower case light face italic Latin letters (e.g. f and h) are used to label scalars.
   Vectors: non-indexed lower or upper case bold face non-italic Latin letters (e.g. **a** and **A**) are used to label vectors in symbolic notation with the exception of the basis vectors where indexed bold face lower or upper case non-italic symbols (e.g. $\mathbf{e}_1$ and $\mathbf{E}^i$) are used.
   Tensors of rank > 1: non-indexed upper case bold face non-italic Latin letters (e.g. **A** and **B**) are used to label tensors of rank > 1 in symbolic notation.
   Indexed light face italic Latin symbols (e.g. a_i and B_i^{jk}) are used to denote tensors of rank > 0 (i.e. vectors and tensors of rank > 1) in their explicit tensor form, i.e. index notation.

2. What the comma and semicolon in $A^{jk}_{,i}$ and $A_{k;i}$ mean?
   Answer: The comma means partial derivative with respect to the variable whose index follows the comma (i.e. the ith variable in $A^{jk}_{,i}$), while semicolon means covariant derivative with respect to the variable whose index follows the semicolon (i.e. the ith variable in $A_{k;i}$).

3. State the summation convention and explain its conditions. To what type of indices this convention applies?
   Answer: According to the summation convention, dummy indices imply summation over their range. More clearly, a twice-repeated variable (i.e. not numeric) index in a single term (whether the twice-repeated index occurs in one tensor or in two tensors) implies a sum of terms equal in number to the range of the repeated index. Hence, in a 2D space we have:
   $$B^a_a = B^1_1 + B^2_2$$
   while in a 3D space we have:
   $$C_a D^a = C_1 D^1 + C_2 D^2 + C_3 D^3$$

4. What is the number of components of a rank-3 tensor in a 4D space?
   Answer: The number of components of a rank-r tensor in an nD space is given by n^r. Hence, the number of components is $4^3 = 64$.

---

[1] This answer is about the symbolism of this book (which is generally of common use), and hence some conditions (e.g. being of lower or upper case) are not universal. The readers should therefore consult each author about his own convention about these conditions.

1 PRELIMINARIES

5. A symbol like B_i^{jk} may be used to represent tensor or its components. What is the difference between these two representations? Do the rules of indices apply to both representations or not? Justify your answer.
   Answer: The difference is that when these symbols represent tensors they should be treated as tensors and hence they obey the rules of tensors (e.g. the transformation rules and the rules of indices), while when they represent components they are like scalars and hence they are ordinary numbers or variables (e.g. real numbers). For example, when B_i^{jk} represents a tensor it is wrong to write $B_i^{jk} + C$ where C is a scalar, but this is correct when B_i^{jk} represents a component.
   As indicated, the rules of indices apply to tensors but not to their components.

6. What is the meaning of the following symbols: ∇, ∂_j, ∂_{kk}, ∇^2, ∂_ϕ, $h_{,jk}$, $A^i_{;n}$, ∂^n, $\nabla^{;k}$ and $C_{i;km}$?
   Answer:
   ∇: nabla differential vector operator.
   ∂_j: partial derivative operator with respect to the jth variable.
   ∂_{kk}: Laplacian differential scalar operator in Cartesian form.
   ∇^2: Laplacian differential scalar operator.
   ∂_ϕ: partial derivative with respect to the variable ϕ.
   $h_{,jk}$: second order partial derivative with respect to the variables indexed by j and k.
   $A^i_{;n}$: covariant derivative of the contravariant vector A^i with respect to the variable indexed by n.
   ∂^n: contravariant derivative with respect to the variable indexed by n.
   $\nabla^{;k}$: contravariant derivative with respect to the variable indexed by k.
   $C_{i;km}$: second order covariant derivative of the covariant vector C_i with respect to the variables indexed by k and m.

7. What is the difference between symbolic notation and indicial notation? For what type of tensors these notations are used? What are the other names given to these types of notation?
   Answer: The symbolic notation is a geometrically oriented notation with no reference to a particular coordinate system and hence it is intrinsically invariant to the choice of coordinate system, while the indicial notation takes an algebraic form based on components identified by indices and referred to a particular set of basis vectors of a given coordinate system and hence the notation is suggestive of an underlying coordinate system. Also, the symbolic notation is usually identified by using bold face non-italic symbols, like **a** and **B**, while the indicial notation is identified by using light face indexed italic symbols such as a^i and B_{ij}.
   These notations are used for non-scalar tensors and hence they belong to tensors of rank > 0.
   Other names for symbolic notation are index-free notation, direct notation, and Gibbs notation.
   Other names for indicial notation are index notation, component notation, and tensor notation.

8. "The characteristic property of tensors is that they satisfy the principle of invariance

under certain coordinate transformations". Does this mean that the components of tensors are constant? Why this principle is very important in physical sciences?

Answer: No. The principle of invariance is about the invariance of the form and not about the invariance or constancy of the values of the individual components.[2]

This principle is very important in physical sciences because the laws of science should satisfy the principle of form-invariance when they are transformed across coordinate systems and frames of reference. This is because for the laws of science to be useful and of common value, they should be independent of the coordinate systems, frames of reference and observers.

9. State and explain all the notations used to represent tensors of all ranks (rank-0, rank-1, rank-2, etc.). What are the advantages and disadvantages of using each one of these notations?

    Answer: Regarding rank-0 tensors (i.e. scalars), they have only one way of labeling which is commonly non-indexed light face italic Latin letters (e.g. f) or Greek letters (e.g. ϕ).

    Regarding tensors of rank > 0 (i.e. vectors and higher rank tensors), they have symbolic notation and indicial notation which are explained in a previous question (see Exercise 7). The symbolic notation is of geometric nature with no reference to a particular coordinate system, while the indicial notation is of algebraic nature with an indication to an underlying coordinate system and basis tensors. Non-indexed bold face straight symbols are usually used to represent symbolic notation, while indexed light face italic symbols are usually used to represent indicial notation. Symbolic notation is used in general representation while indicial notation is used in specific representations, formulations and calculations.

    Regarding the advantages and disadvantages, symbolic notation is more general and succinct and easier to read than indicial notation, while indicial notation is more specific and informative. Indicial notation may be susceptible to some confusion since the same symbol (like A_i) may be used to represent the components as well as the tensor itself but from this very perspective it is more flexible and versatile. Indicial notation may also be more susceptible to error in writing and typesetting due to the presence of indices and may also require more overhead in this regard since writing and typesetting indexed symbols in a legible form usually require extra effort especially when using simple editors, for example, although bold-facing (or using similar notational techniques like underlining or using over-arrows) also requires additional effort. In brief, symbolic notation is recommended for general representation while indicial notation should be used in specific representation that requires the revelation of the underlying structure and indication of the coordinate system or reference frame and basis vectors such as

---

[2] This also implies the invariance of what the tensor represents of abstract mathematical entity or physical entity and hence the "reality" of the represented entity is independent of the form and type of representation. For example, a vector of magnitude 1 meter pointing to the north will be so in any coordinate system and in any form of representation and hence it will not be of magnitude 1 in one system and of magnitude 2 in another system or pointing north in one form of representation and pointing east in another form of representation.

during formulation and calculation. For example, when we talk about a tensor that can be covariant or contravariant or mixed or we want to talk about a tensor whose variance type is irrelevant in that context, then it is more appropriate to use symbolic notation like **A** because it is general and can represent any variance type, but if we talk specifically about the properties and the rules that apply specifically to one of these variance types (such as covariant type) or we intend to use the tensor symbol in explicit tensor formulation, calculation and development of analytical arguments and proofs then it is more appropriate (and may even be necessary) to use indicial notation like A_i for that tensor.

10. State the continuity condition that should be met if the equality: $\partial_i \partial_j = \partial_j \partial_i$ is to be correct.
    Answer: The continuity condition means that the function and its first and second partial derivatives do exist and they are continuous in their domain.

11. Explain the difference between free and bound tensor indices. Also, state the rules that govern each one of these types of index in tensor terms, expressions and equalities.
    Answer: The differences can be summarized as follows:
    (a) Free index occurs only once in a tensor term, while bound (or dummy) index occurs twice in a tensor term.
    (b) The summation convention applies to bound indices but not to free indices.
    (c) Free indices have extended presence in all terms of tensor expressions and equalities, while bound indices are restricted to their terms and hence they can occur only in some terms of tensor expressions and equalities.
    (d) Bound indices can be replaced in individual terms (as long as the new label is not used in that term) but free indices cannot although free indices can be replaced in all terms if the new label is not in use in that context.
    (e) When bound indices are present in more than one term of tensor expressions and equalities they can be named differently in each term but free indices should be named uniformly in all terms.
    (f) Free indices count in tensor rank and order but bound indices count only in tensor order.
    (g) Bound indices can be present in scalar quantity (when all indices are contracted) but free indices can not.
    The rules of free indices:
    (A) Each term should have the same number of free indices.
    (B) A free index should have the same variance type in all terms.
    (C) Each term should have the same set of free indices, e.g. all terms should have i, j, k and hence it is not allowed to have one term with i, j, k set and another term with i, j, n set.
    (D) The free indices should have the same arrangement in all terms, e.g. $ijkl$ in all terms and not $ijkl$ in some terms and $ikjl$ in other terms.
    (E) Each index should have the same range (i.e. space dimensionality) in all terms, and hence the index i in $A_i + B_i$ expression should have identical range in both terms.
    The rules of bound indices:

(i) They are usually subject to the summation convention.
(ii) They generally should be of opposite variance type (i.e. one covariant and one contravariant) except in orthonormal Cartesian systems where they can have the same variance type.
(iii) They can be named independently in each term.
(iv) They do not contribute to the tensor rank but they contribute to the tensor order.

12. Explain the difference between the order and the rank of tensors and link this to the free and dummy indices.
    Answer: The order represents the total number of indices including dummy indices, while the rank represents the number of free indices only. Accordingly, free indices count in tensor rank and order but dummy indices count only in tensor order.

13. What is the difference between covariant, contravariant and mixed type tensors? Give examples for each.
    Answer: In brief:
    (a) Covariant tensors have only subscript indices. Contravariant tensors have only superscript indices. Mixed tensors have both subscript and superscript indices.
    (b) Covariant and contravariant tensors are of rank > 0 (i.e. vectors and higher rank tensors and hence they exclude only scalars), while mixed tensors are of rank > 1 (and hence they exclude scalars and vectors).
    (c) Covariant tensors are associated with contravariant basis vectors. Contravariant tensors are associated with covariant basis vectors. Mixed tensors are associated with both covariant and contravariant basis vectors (i.e. their covariant/contravariant indices correspond to contravariant/covariant basis vectors).
    Examples: A_i, B_{ij}, C_{ijk} are covariant tensors. A^i, B^{ij}, C^{ijk} are contravariant tensors. A^i_j, B^{ij}_k, C^{mn}_{ijk} are mixed tensors.

14. What is the meaning of "unit" and "zero" tensors? What is the characteristic feature of these tensors with regard to the value of their components?
    Answer: Unit tensor is a tensor whose all components are zero except those with identical values of all indices which are assigned the value 1. Zero tensor is a tensor whose all components are zero. The characteristic feature of these tensors is that all their components are constant (i.e. 0 and 1 for unit tensor and 0 for zero tensor).

15. What is the meaning of "orthonormal vector set" and "orthonormal coordinate system"? State any relevant mathematical condition.
    Answer: Orthonormal vector set means a set of vectors which are mutually orthogonal and each one is of unit length. The orthonormality of a vector set may be expressed mathematically by the following dot product equations:
    $$\mathbf{V}_i \cdot \mathbf{V}_j = \delta_{ij} \qquad \text{or} \qquad \mathbf{V}^i \cdot \mathbf{V}^j = \delta^{ij}$$
    where the indexed δ is the Kronecker delta symbol and the indexed $\mathbf{V}$ symbolizes a vector in the set while i and j are ranging over the dimension of the underlying space. Orthonormal coordinate system means a coordinate system whose basis vector set is orthonormal at all points of the space where the system is defined.

16. What is the rule that governs the pair of dummy indices involved in summation re-

garding their variance type in general coordinate systems? Which type of coordinate system is exempt of this rule and why?

Answer: The rule is that the indices should be of opposite variance type (i.e. one covariant and the other contravariant). The coordinate system that is exempt from this rule is the orthonormal Cartesian because the covariant and contravariant types of this system are identical.

17. State all the rules that govern the indicial structure of tensors involved in tensor expressions and equalities (rank, set of free indices, variance type, order of indices and labeling).

    Answer: The rules are:

    (a) Each term should have the same number of free indices, i.e. the same rank.

    (b) A free index should have the same variance type in all terms.

    (c) Dummy indices should be of opposite variance type except in orthonormal Cartesian systems.

    (d) Each term should have the same set of free indices, e.g. all terms should have i, j, k and hence it is not allowed to have one term with i, j, k set and another term with i, j, n set.

    (e) The free indices should have the same arrangement, e.g. i first k second and m third.

    (f) Each index should have the same range (i.e. space dimensionality) in all terms.

    (g) Dummy and free indices should have distinct labeling, i.e. no dummy index can share the same label with a free index.

18. How many equalities that the following equation contains assuming a 4D space: $B_i^k = C_i^k$? Write all these equalities explicitly, i.e. $B_1^1 = C_1^1$, $B_1^2 = C_1^2$, etc.

    Answer: Sixteen. They are:

    $$\begin{array}{llll} B_1^1 = C_1^1 & B_2^1 = C_2^1 & B_3^1 = C_3^1 & B_4^1 = C_4^1 \\ B_1^2 = C_1^2 & B_2^2 = C_2^2 & B_3^2 = C_3^2 & B_4^2 = C_4^2 \\ B_1^3 = C_1^3 & B_2^3 = C_2^3 & B_3^3 = C_3^3 & B_4^3 = C_4^3 \\ B_1^4 = C_1^4 & B_2^4 = C_2^4 & B_3^4 = C_3^4 & B_4^4 = C_4^4 \end{array}$$

19. Which of the following tensor expressions is legitimate and which is not, giving detailed explanation in each case?

    $$A_i^k - B_i, \qquad C_a^a + D_m^n - B_b^b, \qquad a + B, \qquad S_{cdk}^{cdj} + F_{abk}^{abj}$$

    Answer:
    First: illegitimate because the number of free indices is different, i.e. the two tensors are of different rank.

Second: illegitimate because the involved tensors are of different rank, i.e. C_a^a and B_b^b are of rank-0 while D_m^n is of rank-2.

Third: legitimate because a and B are scalars.

Fourth: legitimate because S_{cdk}^{cdj} and F_{abk}^{abj} have identical indicial structure that follows all the rules of free and bound indices in tensor expressions. The difference in the dummy indices (i.e. c and d in S_{cdk}^{cdj} and a and b in F_{abk}^{abj}) does not matter because bound indices are restricted to their own term.

20. Which of the following tensor equalities is legitimate and which is not, giving detailed explanation in each case?

$$A_i^{\;n} = B_{.i}^{n.}, \qquad D = S_c^c + N_{ba}^{ab}, \qquad 3a + 2b = J_a^a, \qquad B_k^m = C_m^k, \qquad B_j = 3c - D_j$$

Answer:
First: illegitimate because the arrangement of the indices is different, i.e. in in $A_i^{\;n}$ and ni in $B_{.i}^{n.}$.

Second: legitimate because all the involved tensors in this equality are of rank-0.

Third: legitimate because all the involved tensors are of rank-0.

Fourth: illegitimate because the corresponding indices are of different variance type.

Fifth: illegitimate because the involved tensors are of different rank, i.e. B_j and D_j are of rank-1 while c is of rank-0.

21. Explain why the indicial structure (rank, set of free indices, variance type and order of indices) of tensors involved in tensor expressions and equalities are important referring in your explanation to the vector basis set to which the tensors are referred. Also explain why these rules are not observed in the expressions and equalities of tensor components.

Answer: The reason is that the indicial notation of tensors is based on a particular set of basis vectors and hence the characteristics of the indicial structure (i.e. rank, set of free indices, variance type and order of indices) have particular significance since they have certain association and representation of the basis set and its characteristics. For example, the tensor $A_i^{\;j}$ means $A_i^{\;j}\mathbf{E}^i\mathbf{E}_j$ while the tensor $B^i_{\;jk}$ means $B^i_{\;jk}\mathbf{E}_i\mathbf{E}^j\mathbf{E}^k$ and hence $A_i^{\;j} + B^i_{\;jk}$ and $A_i^{\;j} = B^i_{\;jk}$ are incorrect because $A_i^{\;j}\mathbf{E}^i\mathbf{E}_j + B^i_{\;jk}\mathbf{E}_i\mathbf{E}^j\mathbf{E}^k$ and $A_i^{\;j}\mathbf{E}^i\mathbf{E}_j = B^i_{\;jk}\mathbf{E}_i\mathbf{E}^j\mathbf{E}^k$ are meaningless.

The rules of indicial structure are not observed in the expressions and equalities of tensor components because these components are scalars in nature and hence these expressions and equalities do not refer to the basis vectors. For example, if the value of the component B_{ij} of the tensor $\mathbf{B}$ is 10 and the value of the component C^{ij} of the tensor $\mathbf{C}$ is also 10 then it is meaningful and useful to state $B_{ij} = C^{ij}$. Similarly, when we write $\epsilon_{ij} = \epsilon^{ij}$ we mean the corresponding components of the tensors ϵ_{ij} and ϵ^{ij} have identical values, e.g. $\epsilon_{12} = \epsilon^{12} = 1$ and $\epsilon_{11} = \epsilon^{11} = 0$ which is legitimate and correct.

22. Why free indices should be named uniformly in all terms of tensor expressions and equalities while dummy indices can be named in each term independently?

Answer: The reason is that free indices refer to the common basis vector set and hence they have reach beyond their own individual terms (i.e. they represent a common reference across all terms of the tensor expressions and equalities), while each dummy

1 PRELIMINARIES

index represents a sum in its own term with no reach or presence into other terms and hence dummy indices can be named independently in each term.

23. What are the rules that should be observed when replacing the symbol of a free index with another symbol? What about replacing the symbols of dummy indices?
    Answer: The rules about replacing the symbol of a free index are:
    (a) The replacement should be uniform and thorough within the given context and hence it should take place over all the occurrences of the replaced index in that context and not only over some terms or expressions or equalities.
    (b) The new index should not be in use already as a label to another free or bound index in that context.
    (c) All the indicial structural aspects of the old index (variance type, position, etc.) should be inherited by the new index. In brief, the change should be restricted to naming and should not touch any other aspect of the replaced index.
    Regarding the replacement of the symbols of dummy indices, they can be replaced by another symbol which is not present as a free or dummy index in their term as long as there is no confusion with a similar symbol in that context.

24. Why in general we have: $\partial_i A_j \neq A_j \partial_i$? What are the situations under which the following equality is valid: $\partial_i A_j = A_j \partial_i$?
    Answer: The reason is that $\partial_i A_j$ and $A_j \partial_i$ have different meaning and mathematical significance and implication because $\partial_i A_j$ means that ∂_i is operating on A_j while $A_j \partial_i$ means that ∂_i is operating on something else and A_j just multiplies the result of this operation.
    The equality $\partial_i A_j = A_j \partial_i$ holds identically when $A_j = 0$. It also holds in many other special cases. For example, it holds when A_j and the operand of ∂_i on the right hand side are constant (or more generally they are independent of the ith variable and hence the partial derivatives are zero).

25. What is the difference between the order of a tensor and the order of its indices?
    Answer: The order of a tensor is an indicator of the total number of tensor indices, while the order of its indices represents the arrangement of these indices (i.e. which is first, which is second, etc.).

26. In which case A_{ijk} is equal to A_{ikj}? What about A_{ijk} and A^{ikj}?
    Answer: A_{ijk} is equal to A_{ikj} when we have a symmetry in the j and k indices.
    A_{ijk} and A^{ikj} are equal when we have an underlying orthonormal Cartesian coordinate system plus a symmetry in the j and k indices.[3]

27. What are the rank, order and dimension of the tensor A^i_{jk} in a 3D space? What about the scalar f and the tensor A^{abm}_{abjn} from the same perspectives?
    Answer: The rank of A^i_{jk} is 3, its order is 3 and its dimension is 3.
    The rank of f is 0, its order is 0 and its dimension is 3.
    The rank of A^{abm}_{abjn} is 3, its order is 7 and its dimension is 3.

28. What is the order of indices in $A_j{}^i{}_k$? Insert a dot in this symbol to make the order

---

[3] We note that the perspective of components and basis vectors my not be entirely consistent in this answer. However, the main focus is the condition for symmetry.

more explicit.

Answer: The order is *jik*. On inserting a dot between j and k we get $A_j{}^i{}_{.\,k}$ which is more explicit about the order of indices.

29. Why the order of indices of mixed tensors may not be clarified by using spaces or inserting dots?

    Answer: Some reasons are:

    • The order of indices is irrelevant in the given context, e.g. any order can achieve the intended purpose.

    • The order is clear from other indicators in the given context.

    • The order is indicated implicitly by the alphabetical order of the selected indices and hence A_i^j for instance means i first and j second.

30. What is the meaning of "tensor field"? Is A^i a tensor field considering the spatial dependency of A^i and the meaning of "tensor"?

    Answer: Tensor field is a tensor that is defined over an extended and continuous region (or regions) of the space or over the whole space.

    Yes, the vector A^i should be a tensor field when it is defined over an extended region of the space where "tensor" here is used in its general sense that includes vectors.

Chapter 2
Spaces, Coordinate Systems and Transformations

1. Give brief definitions to the following terms: Riemannian space, coordinate system and metric tensor.
   Answer:
   Riemannian space is a manifold characterized by the existence of a symmetric rank-2 tensor called the metric tensor that is defined over the whole manifold.
   Coordinate system is an abstract mathematical device of geometric nature that is used by an observer to identify the position of points and objects and describe events in a given space or manifold.
   Metric tensor is a rank-2 symmetric absolute non-singular tensor that is associated with a given Riemannian space. The metric tensor contains vital information about the essential geometric properties of the space
2. Discuss the main functions of the metric tensor in a given space. How many types the metric tensor can have?
   Answer: The functions of the metric tensor include:
   • Identifying the geometric properties of the space.
   • Raising and lowering indices and hence facilitating the transformation between the covariant and contravariant types.
   Types of the metric tensor: covariant, contravariant and mixed.
3. What is the meaning of "flat" and "curved" space? Give mathematical conditions for the space to be flat in terms of the length of an infinitesimal element of arc and in terms of the metric tensor. Why these conditions should be global for the space to be flat?
   Answer: Flat space is a space to which a coordinate system whose metric tensor can be cast into a diagonal form with all the diagonal entries being $+1$ or -1 does exist, while curved space is a space to which such a coordinate system does not exist.
   The mathematical condition for an nD space to be flat in terms of the length of an infinitesimal element of arc ds is given by:

$$(ds)^2 = \zeta_1(du^1)^2 + \zeta_2(du^2)^2 + \ldots + \zeta_n(du^n)^2 = \sum_{i=1}^{n} \zeta_i(du^i)^2$$

where the indexed ζ are ± 1 while the indexed u are the coordinates of the space. The mathematical condition for an nD space to be flat in terms of the metric tensor is given by:

$$g_{ij} = \pm 1 \qquad (i = j)$$

$$g_{ij} = 0 \qquad (i \neq j)$$

where g_{ij} are the elements of the metric tensor.

When we describe a space to be flat we mean globally flat (otherwise we describe it as locally flat) and hence these conditions should be global for the space to be flat in a global sense.

4. Give common examples of flat and curved spaces of different dimensions justifying in each case why the space is flat or curved.
   Answer:
   • Plane is an example of a 2D flat space because it can be coordinated by a 2D Cartesian system with a diagonal metric tensor whose all diagonal elements are $+1$.
   • Ordinary Euclidean space is an example of a 3D flat space because it can be coordinated by a 3D Cartesian system with a diagonal metric tensor whose all diagonal elements are $+1$.
   • Minkowski spacetime manifold that underlies the mechanics of Lorentz transformations is an example of a 4D flat space because it can be coordinated by a 4D coordinate system with a diagonal metric tensor whose all diagonal elements are ± 1.
   • Sphere and ellipsoid are examples of 2D curved space because they cannot be coordinated by a system with a diagonal metric tensor whose diagonal elements are ± 1.
   • Examples of curved spaces of higher dimensionality can be found in mathematics and some theories of modern physics where abstract curved spaces are used to conceptualize and quantify mathematical and physical theories. For example, in the general theory of relativity curved 4D spaces (representing the spacetime of the physical world) are used to formulate a geometric theory of gravity.

5. Explain why all 1D spaces are Euclidean.
   Answer: The reason is that any curve can be mapped isometrically to a straight line where both are naturally parameterized by arc length. This means that any curvature of a 1D space does not belong to the space itself but to the embedding space, i.e. the curvature is extrinsic rather than intrinsic.

6. Give examples of spaces with constant curvature and spaces with variable curvature.
   Answer: Plane, sphere and Beltrami pseudo-sphere are examples of 2D spaces with constant curvature (where the curvature of plane is 0, the curvature of sphere is $\frac{1}{r^2}$ with r being its radius and the curvature of Beltrami pseudo-sphere is $-\frac{1}{\rho^2}$ with ρ being the pseudo-radius of the pseudo-sphere), while ellipsoid, torus and elliptic and hyperbolic paraboloids are examples of 2D spaces with variable curvature.

7. State Schur theorem outlining its significance.
   Answer: Schur theorem in differential geometry asserts that if the Riemann-Christoffel curvature tensor at each point of an nD space ($n > 2$) is a function of the coordinates only, then the curvature is constant all over the space. An example of its practical significance is that by performing a simple test on the dependency of the Riemann-Christoffel curvature tensor and finding the curvature on a single point we will have information about the curvature of the space at a global level. The theory also has other important theoretical significance and implications.

8. What is the condition for a space to be intrinsically flat and extrinsically flat?
   Answer: A space is intrinsically flat *iff* the Riemann-Christoffel curvature tensor vanishes identically over the space, and it is extrinsically (as well as intrinsically) flat *iff* the curvature tensor vanishes identically over the whole space.
9. What is the common method of investigating the Riemannian geometry of a curved manifold?
   Answer: The common method for investigating the Riemannian geometry of a curved manifold is to embed the manifold in a Euclidean space of higher dimensionality and inspect the properties of the manifold from this perspective.
10. Give brief definitions to coordinate curves and coordinate surfaces outlining their relations to the basis vector sets. How many independent coordinate curves and coordinate surfaces we have at each point of a 3D space with a valid coordinate system?
    Answer: Coordinate curves are curves along which exactly one coordinate varies while all the other coordinates are constant, while coordinate surfaces are surfaces over which exactly one coordinate is constant while all the other coordinates vary. The covariant basis vectors are tangent vectors to the coordinate curves, while the contravariant basis vectors are gradient of the space coordinates and hence they are perpendicular to the coordinate surfaces.
    We should have 3 independent coordinate curves and 3 independent coordinate surfaces at each point of a 3D space with a valid coordinate system.
11. Why a coordinate system is needed in tensor formulations?
    Answer: Coordinate systems are needed in tensor calculus to define non-scalar tensors in a specific form and identify their components in reference to the basis set of the system.
12. List the main types of coordinate system outlining their relations to each other.
    Answer: Coordinate systems can be classified from different perspectives.
    • For example, they can be classified as rectilinear coordinate systems which are characterized by the property that all their coordinate curves are straight lines and all their coordinate surfaces are planes, and curvilinear coordinate systems which are characterized by the property that at least some of their coordinate curves are not straight lines and some of their coordinate surfaces are not planes.
    • They can also be classified as orthogonal coordinate systems which are characterized by having mutually perpendicular coordinate curves and coordinate surfaces at each point in their space, and non-orthogonal which are not so.
    • They can also be classified as homogeneous when the metric tensor of their space is the unity tensor, and non-homogeneous otherwise.
    • They may also be classified individually from the perspective of their own characteristics and hence we have Cartesian, cylindrical, spherical as well as many other types of coordinate systems (e.g. parabolic and parabolic cylindrical) whose properties are thoroughly investigated in mathematical texts.
13. "The coordinates of a system can have the same physical dimension or different physical dimensions". Give an example for each.
    Answer: The coordinates of Cartesian systems have the same physical dimension (i.e.

length), while the coordinates of cylindrical systems have different physical dimensions (i.e. length for ρ and z and angle for ϕ which is dimensionless).

14. Prove that spherical coordinate systems are orthogonal.
    Answer: Orthogonal systems are characterized by having mutually perpendicular basis vectors, and hence all we need for establishing this proof is to show that:
    $$\mathbf{e}_r \cdot \mathbf{e}_\theta = 0 \qquad \mathbf{e}_r \cdot \mathbf{e}_\phi = 0 \qquad \mathbf{e}_\theta \cdot \mathbf{e}_\phi = 0$$
    The basis vectors for spherical coordinate systems are given in orthonormal Cartesian form by the following equations:
    $$\begin{aligned}
    \mathbf{e}_r &= \sin\theta\cos\phi\,\mathbf{i} + \sin\theta\sin\phi\,\mathbf{j} + \cos\theta\,\mathbf{k} \\
    \mathbf{e}_\theta &= \cos\theta\cos\phi\,\mathbf{i} + \cos\theta\sin\phi\,\mathbf{j} - \sin\theta\,\mathbf{k} \\
    \mathbf{e}_\phi &= -\sin\phi\,\mathbf{i} + \cos\phi\,\mathbf{j}
    \end{aligned}$$
    where $\mathbf{i}, \mathbf{j}, \mathbf{k}$ are the Cartesian unit basis vectors. Accordingly, we have:
    $$\begin{aligned}
    \mathbf{e}_r \cdot \mathbf{e}_\theta &= \sin\theta\cos\phi\cos\theta\cos\phi + \sin\theta\sin\phi\cos\theta\sin\phi - \cos\theta\sin\theta \\
    &= \sin\theta\cos\theta\cos^2\phi + \sin\theta\cos\theta\sin^2\phi - \cos\theta\sin\theta \\
    &= \sin\theta\cos\theta\left(\cos^2\phi + \sin^2\phi\right) - \cos\theta\sin\theta \\
    &= \sin\theta\cos\theta - \cos\theta\sin\theta \\
    &= 0
    \end{aligned}$$

    $$\mathbf{e}_r \cdot \mathbf{e}_\phi = -\sin\theta\sin\phi\cos\phi + \sin\theta\sin\phi\cos\phi = 0$$

    $$\mathbf{e}_\theta \cdot \mathbf{e}_\phi = -\cos\theta\cos\phi\sin\phi + \cos\theta\cos\phi\sin\phi = 0$$

15. What is the difference between rectilinear and curvilinear coordinate systems?
    Answer: All the coordinate curves of rectilinear coordinate systems are straight lines and all their coordinate surfaces are planes, while the coordinate curves and coordinate surfaces of curvilinear systems are not so and hence at least some of their coordinate curves are not straight lines and some of their coordinate surfaces are not planes. Consequently, the basis vectors of rectilinear systems are constant while the basis vectors of curvilinear systems are variable in general since their direction or/and magnitude depend on the position in the space and hence they are coordinate dependent.

16. Give examples of common curvilinear coordinate systems explaining why they are curvilinear.
    Answer: The most common examples are the cylindrical and spherical coordinate systems.
    The cylindrical coordinate systems are curvilinear because the ρ, ϕ, z coordinate curves are straight lines, circles and straight lines respectively (and hence some of their coordinate curves are not straight lines), while the ρ, ϕ, z coordinate surfaces are cylinders, semi-planes and planes respectively (and hence some of their coordinate surfaces are

not planes).
The spherical coordinate systems are curvilinear because the r, θ, ϕ coordinate curves are straight lines, semi-circles and circles respectively (and hence some of their coordinate curves are not straight lines), while the r, θ, ϕ coordinate surfaces are spheres, cones and semi-planes respectively (and hence some of their coordinate surfaces are not planes).

17. Give an example of a commonly used curvilinear coordinate system with some of its coordinate curves being straight lines.
    Answer: An example is the cylindrical coordinate system whose ρ and z coordinate curves are straight lines.

18. Define briefly the terms "orthogonal" and "homogeneous" coordinate system.
    Answer: Orthogonal coordinate system is a system whose coordinate curves, as well as its coordinate surfaces, are mutually perpendicular at each point in the space. Accordingly, the vectors of its covariant basis set and the vectors of its contravariant basis set are mutually orthogonal everywhere in the space.
    Homogeneous coordinate system is a system associated with the unity tensor as the metric of its underlying space.

19. Give examples of rectilinear and curvilinear orthogonal coordinate systems.
    Answer: Orthonormal Cartesian systems are examples of rectilinear orthogonal coordinate systems, while cylindrical and spherical coordinate systems are examples of curvilinear orthogonal coordinate systems.

20. What is the condition of a coordinate system to be orthogonal in terms of the form of its metric tensor? Explain why this is so.
    Answer: The necessary and sufficient condition for a coordinate system to be orthogonal is that its metric tensor is diagonal. This can be inferred from the definition of the components of the metric tensor as the dot products of the basis vectors since the dot product involving two different vectors (i.e. $\mathbf{E}_i \cdot \mathbf{E}_j$ or $\mathbf{E}^i \cdot \mathbf{E}^j$ with $i \neq j$) will vanish if the basis vectors, whether covariant or contravariant, are mutually perpendicular. As the condition $i \neq j$ is associated with the non-diagonal components of the metric tensor then this means that all the non-diagonal components are zero and hence the tensor is diagonal.

21. What is the mathematical condition for a coordinate system to be homogeneous?
    Answer: The mathematical condition for a coordinate system to be homogeneous may be given in terms of the metric tensor that associates the system by:

    $$g_{ij} = +1 \qquad (i = j)$$
    $$g_{ij} = 0 \qquad (i \neq j)$$

    where g_{ij} are the elements of the metric tensor.

22. How can we homogenize a non-homogeneous coordinate system of a flat space?
    Answer: A coordinate system of a flat space can be homogenized by allowing the coordinates to be imaginary. This is done by redefining the coordinates as:

    $$U^i = \sqrt{\zeta_i} u^i \qquad \text{(no sum over } i\text{)}$$

where $\zeta_i = \pm 1$. The new coordinates U^i are real when $\zeta_i = 1$ and imaginary when $\zeta_i = -1$.

23. Give examples of homogeneous and non-homogeneous coordinate systems.
    Answer: Orthonormal Cartesian systems are examples of homogeneous coordinate systems, while cylindrical and spherical systems are examples of non-homogeneous coordinate systems.

24. Give an example of a non-homogeneous coordinate system that can be homogenized.
    Answer: The coordinate system of the Minkowski spacetime (which is the space of the mechanics of Lorentz transformations whose metric may be given by $\text{diag}\,[-1,+1,+1,+1]$ or $\text{diag}\,[+1,-1,-1,-1]$) is an example of a non-homogeneous coordinate system that can be homogenized by allowing the temporal coordinate (for the first form of the metric) or the spatial coordinates (for the second form of the metric) to be imaginary.

25. Describe briefly the transformation of spaces and coordinate systems stating relevant mathematical relations.
    Answer: A transformation from an nD space to another nD space is a correlation that maps a point from the first space (original) to a point in the second space (transformed) where each point in the original and transformed spaces is identified by n independent coordinates. The transformation of coordinates may be expressed mathematically by the following relation:
    $$\bar{u}^i = \bar{u}^i(u^1, u^2, \ldots, u^n)$$
    where the unbarred and barred indexed u represent the coordinates of the original and transformed spaces and $i = 1, 2, \ldots, n$ with n being the dimension of the spaces.

26. What "injective transformation" means? Is it necessary that such a transformation has an inverse?
    Answer: Injective transformation means one-to-one. It is not necessary that such a transformation has an inverse unless it is surjective (i.e. onto) as well.

27. Write the Jacobian matrix $\mathbf{J}$ of a transformation between two nD spaces whose coordinates are labeled as u^i and $\bar{u}^i$ where $i = 1, \cdots, n$.
    Answer: The Jacobian matrix of such a transformation is given by:
    $$\mathbf{J} = \begin{bmatrix} \frac{\partial u^1}{\partial \bar{u}^1} & \frac{\partial u^1}{\partial \bar{u}^2} & \cdots & \frac{\partial u^1}{\partial \bar{u}^n} \\ \frac{\partial u^2}{\partial \bar{u}^1} & \frac{\partial u^2}{\partial \bar{u}^2} & \cdots & \frac{\partial u^2}{\partial \bar{u}^n} \\ \vdots & \vdots & \ddots & \vdots \\ \frac{\partial u^n}{\partial \bar{u}^1} & \frac{\partial u^n}{\partial \bar{u}^2} & \cdots & \frac{\partial u^n}{\partial \bar{u}^n} \end{bmatrix}$$

28. State the pattern of the row and column indices of the Jacobian matrix in relation to the indices of the coordinates of the two spaces.
    Answer: The pattern is that the indices of u in the numerator provide the indices for the rows while the indices of $\bar{u}$ in the denominator provide the indices for the columns. This indexing pattern may be interchanged.

29. What is the difference between the Jacobian matrix and the Jacobian and what is the relation between them?
    Answer: The difference is that the Jacobian matrix is a matrix while the Jacobian

is a determinant. The relation between the Jacobian matrix and the Jacobian is that the Jacobian is the determinant of the Jacobian matrix and this can be expressed mathematically as:
$$J = \det(\mathbf{J})$$
where J and $\mathbf{J}$ are the Jacobian and the Jacobian matrix respectively.

30. What is the relation between the Jacobian of a given transformation and the Jacobian of its inverse? Write a mathematical formula representing this relation.
    Answer: The Jacobian of the inverse transformation is the reciprocal of the Jacobian of the original transformation. This relation can be expressed mathematically as:
    $$\bar{J} = \frac{1}{J}$$
    where $\bar{J}$ is the Jacobian of the inverse transformation and J is the Jacobian of the original transformation.

31. Is the labeling of two coordinate systems (e.g. barred and unbarred) involved in a transformation relation essential or arbitrary? Hence, discuss if the labeling of coordinates in the Jacobian matrix can be interchanged.
    Answer: Labeling one coordinate system as barred and the other as unbarred is a matter of choice and convenience and hence it is rather arbitrary. Accordingly, the labeling of coordinates in the Jacobian matrix as barred and unbarred can be interchanged and consequently the Jacobian may be notated as unbarred over barred or barred over unbarred. However, although this is a pure notational matter the arbitrariness sometimes propagates even to the terminology where the "Jacobian" is used in reference to the opposite transformation. Therefore, instead of having "Jacobian" and "inverse Jacobian" we prefer to have "Jacobian of the original transformation" and "Jacobian of the inverse transformation".

32. Using the transformation equations between the Cartesian and cylindrical coordinate systems, find the Jacobian matrix of the transformation between these systems, i.e. Cartesian to cylindrical and cylindrical to Cartesian.
    Answer: The equations of coordinate transformation between the Cartesian and cylindrical systems are given by:

    Cartesian to cylindrical: $\quad \rho = \sqrt{x^2 + y^2} \qquad \phi = \arctan(y/x) \qquad z = z$

    Cylindrical to Cartesian: $\quad x = \rho \cos\phi \qquad y = \rho \sin\phi \qquad z = z$

    For the Jacobian matrix of the transformation from Cartesian to cylindrical we have:[4]

    $\frac{\partial \rho}{\partial x} = \frac{x}{\sqrt{x^2+y^2}}$ $\qquad$ $\frac{\partial \rho}{\partial y} = \frac{y}{\sqrt{x^2+y^2}}$ $\qquad$ $\frac{\partial \rho}{\partial z} = 0$

    $\frac{\partial \phi}{\partial x} = \frac{-y}{x^2+y^2}$ $\qquad$ $\frac{\partial \phi}{\partial y} = \frac{x}{x^2+y^2}$ $\qquad$ $\frac{\partial \phi}{\partial z} = 0$

    $\frac{\partial z}{\partial x} = 0$ $\qquad$ $\frac{\partial z}{\partial y} = 0$ $\qquad$ $\frac{\partial z}{\partial z} = 1$

---

[4] When we say "transformation from Cartesian to cylindrical" we mean using the Cartesian to cylindrical transformation equations. This similarly applies to the following transformations.

Therefore, the Jacobian matrix for this transformation is:

$$\mathbf{J} = \begin{bmatrix} \frac{x}{\sqrt{x^2+y^2}} & \frac{y}{\sqrt{x^2+y^2}} & 0 \\ \frac{-y}{x^2+y^2} & \frac{x}{x^2+y^2} & 0 \\ 0 & 0 & 1 \end{bmatrix}$$

For the Jacobian matrix of the transformation from cylindrical to Cartesian we have:

$$\frac{\partial x}{\partial \rho} = \cos\phi \qquad \frac{\partial x}{\partial \phi} = -\rho\sin\phi \qquad \frac{\partial x}{\partial z} = 0$$

$$\frac{\partial y}{\partial \rho} = \sin\phi \qquad \frac{\partial y}{\partial \phi} = \rho\cos\phi \qquad \frac{\partial y}{\partial z} = 0$$

$$\frac{\partial z}{\partial \rho} = 0 \qquad \frac{\partial z}{\partial \phi} = 0 \qquad \frac{\partial z}{\partial z} = 1$$

Therefore, the Jacobian matrix for this transformation is:

$$\mathbf{J} = \begin{bmatrix} \cos\phi & -\rho\sin\phi & 0 \\ \sin\phi & \rho\cos\phi & 0 \\ 0 & 0 & 1 \end{bmatrix}$$

Note: as verification test, we calculate in the following the Jacobians of the two transformations.

The Jacobian of the transformation from Cartesian to cylindrical is:

$$\begin{aligned} J(x,y,z \to \rho,\phi,z) &= \frac{x}{\sqrt{x^2+y^2}} \times \frac{x}{x^2+y^2} - \frac{y}{\sqrt{x^2+y^2}} \times \frac{-y}{x^2+y^2} \\ &= \frac{x^2+y^2}{\sqrt{x^2+y^2}\,(x^2+y^2)} \\ &= \frac{1}{\sqrt{x^2+y^2}} \\ &= \frac{1}{\rho} \end{aligned}$$

while the Jacobian of the transformation from cylindrical to Cartesian is:

$$J(\rho,\phi,z \to x,y,z) = \rho\cos^2\phi + \rho\sin^2\phi = \rho$$

Hence:

$$J(x,y,z \to \rho,\phi,z) = \frac{1}{J(\rho,\phi,z \to x,y,z)}$$

as it should be.

33. Repeat question 32 for the spherical, instead of cylindrical, system to find the Jacobian this time.

Answer: The equations of coordinate transformation between the Cartesian and spherical systems are given by:

Cartesian to spherical:

$$r = \sqrt{x^2+y^2+z^2} \qquad \theta = \arccos\left(\frac{z}{\sqrt{x^2+y^2+z^2}}\right) \qquad \phi = \arctan\left(\frac{y}{x}\right)$$

2 SPACES, COORDINATE SYSTEMS AND TRANSFORMATIONS

Spherical to Cartesian:

$$x = r\sin\theta\cos\phi \qquad y = r\sin\theta\sin\phi \qquad z = r\cos\theta$$

For the Jacobian matrix of the transformation from Cartesian to spherical we have:

$$\frac{\partial r}{\partial x} = \frac{x}{\sqrt{x^2+y^2+z^2}} \qquad \frac{\partial r}{\partial y} = \frac{y}{\sqrt{x^2+y^2+z^2}} \qquad \frac{\partial r}{\partial z} = \frac{z}{\sqrt{x^2+y^2+z^2}}$$

$$\frac{\partial \theta}{\partial x} = \frac{zx}{(x^2+y^2+z^2)\sqrt{x^2+y^2}} \qquad \frac{\partial \theta}{\partial y} = \frac{zy}{(x^2+y^2+z^2)\sqrt{x^2+y^2}} \qquad \frac{\partial \theta}{\partial z} = -\frac{\sqrt{x^2+y^2}}{x^2+y^2+z^2}$$

$$\frac{\partial \phi}{\partial x} = \frac{-y}{x^2+y^2} \qquad \frac{\partial \phi}{\partial y} = \frac{x}{x^2+y^2} \qquad \frac{\partial \phi}{\partial z} = 0$$

Therefore, the Jacobian for this transformation is:[5]

$$J(x,y,z \to r,\theta,\phi) = \begin{vmatrix} \frac{x}{r} & \frac{y}{r} & \frac{z}{r} \\ \frac{zx}{r^2\sqrt{x^2+y^2}} & \frac{zy}{r^2\sqrt{x^2+y^2}} & -\frac{\sqrt{x^2+y^2}}{r^2} \\ \frac{-y}{x^2+y^2} & \frac{x}{x^2+y^2} & 0 \end{vmatrix}$$

$$= \frac{x}{r}\left(\frac{\sqrt{x^2+y^2}}{r^2}\frac{x}{(x^2+y^2)}\right) +$$

$$\frac{y}{r}\left(\frac{\sqrt{x^2+y^2}}{r^2}\frac{y}{(x^2+y^2)}\right) +$$

$$\frac{z}{r}\left(\frac{zx}{r^2\sqrt{x^2+y^2}}\frac{x}{(x^2+y^2)} + \frac{zy}{r^2\sqrt{x^2+y^2}}\frac{y}{(x^2+y^2)}\right)$$

$$= \frac{x^2}{r^3\sqrt{x^2+y^2}} + \frac{y^2}{r^3\sqrt{x^2+y^2}} + \frac{z^2x^2+z^2y^2}{r^3\sqrt{x^2+y^2}(x^2+y^2)}$$

$$= \frac{x^2}{r^3\sqrt{x^2+y^2}} + \frac{y^2}{r^3\sqrt{x^2+y^2}} + \frac{z^2}{r^3\sqrt{x^2+y^2}}$$

$$= \frac{x^2+y^2+z^2}{r^3\sqrt{x^2+y^2}}$$

$$= \frac{r^2}{r^3\sqrt{x^2+y^2}}$$

$$= \frac{1}{r\sqrt{x^2+y^2}}$$

$$= \frac{1}{r^2\left(\sqrt{x^2+y^2}/r\right)}$$

$$= \frac{1}{r^2\sin\theta}$$

---

[5] To be concise, we replace $x^2+y^2+z^2$ with r^2.

For the Jacobian matrix of the transformation from spherical to Cartesian we have:

$\frac{\partial x}{\partial r} = \sin\theta \cos\phi$ $\quad\quad \frac{\partial x}{\partial \theta} = r\cos\theta\cos\phi$ $\quad\quad \frac{\partial x}{\partial \phi} = -r\sin\theta\sin\phi$

$\frac{\partial y}{\partial r} = \sin\theta \sin\phi$ $\quad\quad \frac{\partial y}{\partial \theta} = r\cos\theta\sin\phi$ $\quad\quad \frac{\partial y}{\partial \phi} = r\sin\theta\cos\phi$

$\frac{\partial z}{\partial r} = \cos\theta$ $\quad\quad \frac{\partial z}{\partial \theta} = -r\sin\theta$ $\quad\quad \frac{\partial z}{\partial \phi} = 0$

Therefore, the Jacobian for this transformation is:

$$
\begin{aligned}
J(r,\theta,\phi \to x,y,z) &= \begin{vmatrix} \sin\theta\cos\phi & r\cos\theta\cos\phi & -r\sin\theta\sin\phi \\ \sin\theta\sin\phi & r\cos\theta\sin\phi & r\sin\theta\cos\phi \\ \cos\theta & -r\sin\theta & 0 \end{vmatrix} \\
&= \sin\theta\cos\phi \left(r\sin\theta\cos\phi\, r\sin\theta\right) + \\
&\quad r\cos\theta\cos\phi\left(r\sin\theta\cos\phi\cos\theta\right) - \\
&\quad r\sin\theta\sin\phi\left(-\sin\theta\sin\phi\, r\sin\theta - r\cos\theta\sin\phi\cos\theta\right) \\
&= r^2\sin^3\theta\cos^2\phi + r^2\cos^2\theta\sin\theta\cos^2\phi + \\
&\quad r^2\sin^3\theta\sin^2\phi + r^2\cos^2\theta\sin\theta\sin^2\phi \\
&= r^2\sin\theta\left(\sin^2\theta\cos^2\phi + \cos^2\theta\cos^2\phi + \sin^2\theta\sin^2\phi + \cos^2\theta\sin^2\phi\right) \\
&= r^2\sin\theta\left(\cos^2\phi\left[\sin^2\theta + \cos^2\theta\right] + \sin^2\phi\left[\sin^2\theta + \cos^2\theta\right]\right) \\
&= r^2\sin\theta\left(\cos^2\phi + \sin^2\phi\right) \\
&= r^2\sin\theta
\end{aligned}
$$

As we see, we have:

$$J(x,y,z \to r,\theta,\phi) = \frac{1}{J(r,\theta,\phi \to x,y,z)}$$

as it should be.

34. Give a simple definition of admissible coordinate transformation.
    Answer: An admissible coordinate transformation is a mapping represented by a sufficiently differentiable set of equations and it is invertible by having a non-vanishing Jacobian.

35. What is the meaning of the C^n continuity condition?
    Answer: The C^n continuity condition means that the function and all its first n partial derivatives do exist and they are continuous over their domain.

36. What "invariant" object or property means? Give some illustrating examples.
    Answer: An object or property is described as invariant if it does not change under certain admissible coordinate transformations. For example, in classical mechanics the value of mass is invariant under the Galilean transformations of time and space coordinates since the value of mass is an intrinsic property of the massive object and hence it is independent of the observer. Similarly, in the mechanics of Lorentz transformations Maxwell's equations are invariant under the Lorentz transformations of spacetime coordinates since the form of these equations does not change under these transformations.

37. What is the meaning of "composition of transformations"? State a mathematical relation representing such a composition.
Answer: Composition of transformations means a succession of transformations where the output of one transformation is taken as an input to the next transformation. This may be expressed mathematically as:
$$T_c(\mathbb{O}) = T_m T_{m-1} \cdots T_2 T_1(\mathbb{O})$$
where the transformations T_i ($i = 1, 2, \cdots, m$) are composed to produce the composite transformation T_c and $\mathbb{O}$ is an object that is transformed by these transformations. In the above equation, the output of the transformation T_1 is taken as an input to the transformation T_2 and so on until finally the output of the transformation T_{m-1} is taken as an input to the transformation T_m to obtain the composite transformation T_c.

38. What is the Jacobian of a composite transformation in terms of the Jacobians of the simple transformations that make the composite transformation? Write a mathematical relation that links all these Jacobians.
Answer: The Jacobian of the composite transformation is the product of the Jacobians of the individual transformations which the composition is made of. This can be expressed mathematically (in reference to the equation of the previous question) as:
$$J_c = J_m J_{m-1} \cdots J_2 J_1$$
where J_c is the Jacobian of the composite transformation T_c and J_i ($i = 1, 2, \cdots, m$) is the Jacobian of the T_i transformation.

39. "The collection of all admissible coordinate transformations with non-vanishing Jacobian form a group". What this means? State your answer in mathematical and descriptive forms.
Answer: This means that they are group in the technical sense of this term according to the group theory, and hence they satisfy the properties of closure, associativity, identity and inverse. Mathematically, if we have a set of transformations $T_1, T_2, \cdots$ defined on a certain domain then they should satisfy the following conditions:
• Closure, i.e. if T_i and T_j are any two transformations in this group then their composition $T_c = T_i T_j$ is also a transformation in the group.
• Associativity, i.e. if T_i, T_j, T_k are any three transformations in this group then we should have:
$$T_i(T_j T_k) = (T_i T_j) T_k$$
• Identity, i.e. there is a single transformation T_I in the group such that:
$$T_I T_m = T_m T_I = T_m$$
where T_m is any transformation in the group.
• Inverse, i.e. for any transformation T_m in the group there is exactly one transformation T_m^{-1} (which is called the inverse of T_m) such that:
$$T_m T_m^{-1} = T_m^{-1} T_m = T_I$$
where T_I is the identity transformation.

40. Is the transformation of coordinates a commutative operation?[6] Justify your answer by an example.
    Answer: The transformation of coordinates is not commutative in general and hence we may have $T_i T_j \neq T_j T_i$. An example of this is the composition of two rotations whose outcome depends on the order of the rotations, as explained and demonstrated in the text.
41. A transformation T_3 with a Jacobian J_3 is a composite transformation, i.e. $T_3 = T_2 T_1$ where the transformations T_1 and T_2 have Jacobians J_1 and J_2. What is the relation between J_1, J_2 and J_3?
    Answer: The relation is:
    $$J_3 = J_2 J_1$$
42. Two transformations, R_1 and R_2, are related by: $R_1 R_2 = I$ where I is the identity transformation. What is the relation between the Jacobians of R_1 and R_2? What we should call these transformations?
    Answer: The Jacobian of the identity transformation is 1. Therefore, the relation between the Jacobians of R_1 and R_2 is:
    $$J_1 J_2 = 1$$
    where J_1 and J_2 are the Jacobians of R_1 and R_2 respectively. This means that J_1 and J_2 are reciprocal of each other. We should call each one of these transformations the inverse of the other transformation.
43. Discuss the transformation of one set of basis vectors of a given coordinate system to another set of opposite variance type of that system and the relation of this to the metric tensor.
    Answer: The covariant basis vectors are transformed to the contravariant basis vectors by the contravariant metric tensor of the system, while the contravariant basis vectors are transformed to the covariant basis vectors by the covariant metric tensor of the system. This can be expressed mathematically as:
    $$\mathbf{E}^i = g^{ij} \mathbf{E}_j \qquad \mathbf{E}_i = g_{ij} \mathbf{E}^j$$
    where $\mathbf{E}_i, \mathbf{E}_j$ are covariant basis vectors, $\mathbf{E}^i, \mathbf{E}^j$ are contravariant basis vectors, g_{ij} is the covariant metric tensor of the system and g^{ij} is the contravariant metric tensor of the system.
44. Discuss the transformation of one set of basis vectors of a given coordinate system to another set of the same variance type of another coordinate system.
    Answer: The transformation of the basis sets of the same variance type between two coordinate systems (unbarred and barred) is given by the following relations:
    $$\mathbf{E}_i = \frac{\partial \bar{u}^j}{\partial u^i} \bar{\mathbf{E}}_j \qquad \bar{\mathbf{E}}_i = \frac{\partial u^j}{\partial \bar{u}^i} \mathbf{E}_j$$

---
[6] More accurately, "commutative" is an attribute to the composition of transformations.

$$\mathbf{E}^i = \frac{\partial u^i}{\partial \bar{u}^j} \bar{\mathbf{E}}^j \qquad\qquad \bar{\mathbf{E}}^i = \frac{\partial \bar{u}^i}{\partial u^j} \mathbf{E}^j$$

where the indexed u and $\bar{u}$ represent the coordinates in the unbarred and barred systems, while the indexed $\mathbf{E}$ and $\bar{\mathbf{E}}$ are the basis vectors of the relevant variance type in the unbarred and barred systems.

45. Discuss and compare the results of question 43 and question 44. Also, compare the mathematical formulation that should apply in each case.
Answer: While in question 43 we are transforming a basis set of the same system from one variance type to another variance type, in question 44 we are transforming a basis set of the same variance type from one system to another system. As we see, the former is facilitated by the metric tensor of the system while the latter is facilitated by the Jacobian matrix between the two systems.

46. Define proper and improper coordinate transformations.
Answer: Proper transformations are those transformations that preserve the handedness (right- or left-handed) of the coordinate system such as rotation, while improper transformations are those transformations that reverse the handedness of the coordinate system such as reflection.

47. What is the difference between positive and negative orthogonal transformations?
Answer: Positive transformations consist solely of translation and rotation while negative transformations include reflection by an odd number of axes reversal. Accordingly, positive transformations can be decomposed into an infinite number of continuously varying infinitesimal positive transformations each one of which emulates an identity transformation while negative transformations cannot.

48. Give detailed definitions of coordinate curves and coordinate surfaces of 3D spaces discussing the relation between them.
Answer: The coordinate curves are the curves along which exactly one coordinate varies while the other coordinates are held constant, while the coordinate surfaces are the surfaces over which all coordinates vary except one which is held constant. Accordingly, the ith coordinate curve is the curve along which only the ith coordinate varies while the ith coordinate surface is the surface over which only the ith coordinate is constant.
In 3D space, the coordinate curves represent the curves of mutual intersection of the coordinate surfaces.

49. For each one of the following coordinate systems, what is the shape of the coordinate curves and coordinate surfaces: Cartesian, cylindrical and spherical?
Answer:
Cartesian: all coordinate curves are straight lines and all coordinate surfaces are planes.
Cylindrical: the ρ, ϕ and z coordinate curves are straight lines, circles, and straight lines respectively, while the ρ, ϕ and z coordinate surfaces are cylinders, semi-planes and planes respectively.
Spherical: the r, θ and ϕ coordinate curves are straight lines, semi-circles, and circles respectively, while the r, θ and ϕ coordinate surfaces are spheres, cones and semi-planes respectively.

50. Make a simple plot representing the ϕ coordinate curve with the ρ and z coordinate surfaces of a cylindrical coordinate system.
   Answer: The plot should look somewhat like Figure 1.

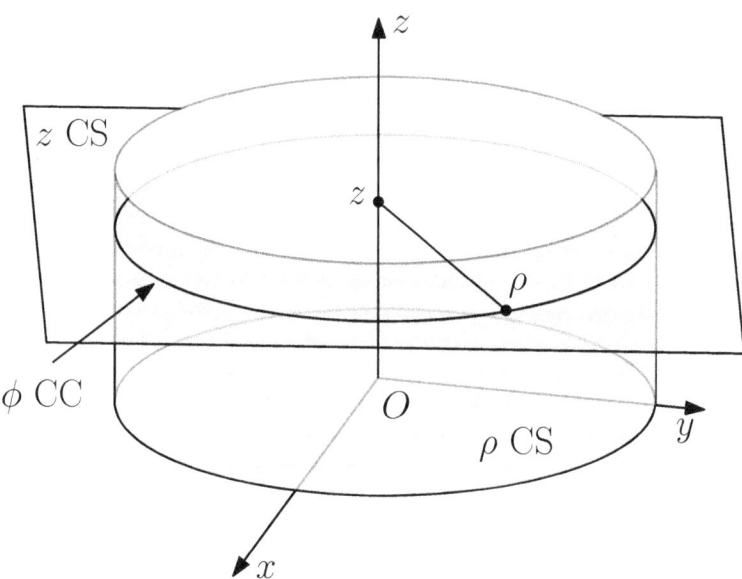

Figure 1: The ϕ coordinate curve (CC) with the ρ and z coordinate surfaces (CS).

51. Make a simple plot representing the r coordinate curve with the θ and ϕ coordinate surfaces of a spherical coordinate system.
   Answer: The plot should look somewhat like Figure 2.

52. Define "scale factors" of a coordinate system and outline their significance.
   Answer: The scale factors of a given coordinate system are factors required to multiply the coordinate differentials to obtain the distances traversed during a change in the coordinate of that magnitude. For example, in cylindrical systems the scale factor ρ is the factor that multiplies the second coordinate ϕ to obtain the distance d traversed in the space by a given change in this coordinate $\Delta\phi$, i.e. $d = \rho\Delta\phi$. Similarly, in spherical systems the scale factor r is the factor that multiplies the second coordinate θ to obtain the distance d traversed in the space by a given change in this coordinate $\Delta\theta$, i.e. $d = r\Delta\theta$. The significance of the scale factors is that they transform the coordinates of the system to lengths which are the real physical dimensions of the space and hence they facilitate the calculation of lengths, areas and volumes as well as any other physical variables that depend on lengths and distances.

53. Give the scale factors of the following coordinate systems: orthonormal Cartesian, cylindrical and spherical.
   Answer:
   Cartesian (x, y, z): $h_x = h_y = h_z = 1$.
   Cylindrical (ρ, ϕ, z): $h_\rho = h_z = 1$ and $h_\phi = \rho$.
   Spherical (r, θ, ϕ): $h_r = 1$, $h_\theta = r$ and $h_\phi = r\sin\theta$.

2 SPACES, COORDINATE SYSTEMS AND TRANSFORMATIONS

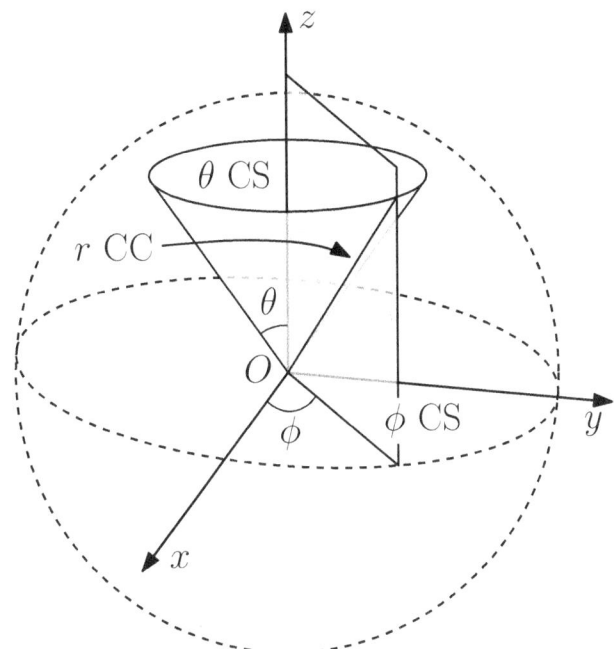

Figure 2: The r coordinate curve (CC) with the θ and ϕ coordinate surfaces (CS).

54. Define, mathematically and in words, the covariant and contravariant basis vector sets explaining any symbols involved in these definitions.
    Answer: The covariant basis vectors are the tangent vectors to the coordinate curves, while the contravariant basis vectors are the gradient of the space coordinates and hence they are perpendicular to the coordinate surfaces. Mathematically, the covariant and contravariant basis vectors are defined respectively by:

    $$\mathbf{E}_i = \frac{\partial \mathbf{r}}{\partial u^i} \qquad\qquad \mathbf{E}^i = \nabla u^i$$

    where $\mathbf{r}$ is the position vector in Cartesian coordinates $(x^1, \ldots, x^n)$, u^i represents general coordinates, and $i = 1, \cdots, n$ with n being the space dimension.

55. What is the relation of the covariant and contravariant basis vector sets with the coordinate curves and coordinate surfaces of a given coordinate system? Make a simple sketch representing this relation for a general curvilinear coordinate system in a 3D space.
    Answer: As stated earlier, the covariant basis vectors are tangents to the coordinate curves, while the contravariant basis vectors are perpendicular to the coordinate surfaces. The sketch should look something like Figure 3.

56. The covariant and contravariant components of vectors can be transformed one to the other. How? State your answer in a mathematical form.
    Answer: The covariant components of a vector $\mathbf{A}$ are obtained from the contravariant components of $\mathbf{A}$ by using the covariant metric tensor g_{ij} (or lowering operator), that

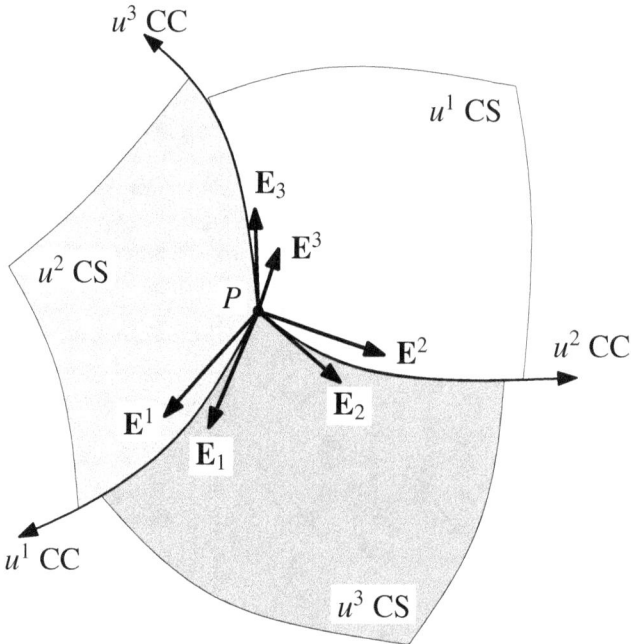

Figure 3: The covariant and contravariant basis vectors of a general curvilinear coordinate system and the associated coordinate curves (CC) and coordinate surfaces (CS) at a given point P in a 3D space.

is:
$$A_i = g_{ij} A^j$$

Similarly, The contravariant components of a vector $\mathbf{B}$ are obtained from the covariant components of $\mathbf{B}$ by using the contravariant metric tensor g^{ij} (or raising operator), that is:
$$B^i = g^{ij} B_j$$

57. What is the significance of the following relations?
$$\mathbf{E}_i \cdot \mathbf{E}^j = \delta_i^j \qquad \mathbf{E}^i \cdot \mathbf{E}_j = \delta^i{}_j$$

   Answer: The significance of these relations is that the covariant and contravariant basis sets are reciprocal basis systems.

58. Write down the mathematical relations that correlate the basis vectors to the components of the metric tensor in their covariant and contravariant forms.
   Answer:
$$\mathbf{E}_i \cdot \mathbf{E}_j = g_{ij} \qquad \mathbf{E}^i \cdot \mathbf{E}^j = g^{ij}$$
   where $\mathbf{E}_i, \mathbf{E}_j$ are covariant basis vectors, $\mathbf{E}^i, \mathbf{E}^j$ are contravariant basis vectors, g_{ij} is the covariant metric tensor and g^{ij} is the contravariant metric tensor.

59. Using the equation $\mathbf{E}_i = \frac{\mathbf{E}^j \times \mathbf{E}^k}{\mathbf{E}^i \cdot (\mathbf{E}^j \times \mathbf{E}^k)}$, show that if $\mathbf{E}^i, \mathbf{E}^j, \mathbf{E}^k$ form a right handed orthonormal system then $\mathbf{E}_i = \mathbf{E}^i$. Repeat the question using this time the equation

$\mathbf{E}^i = \frac{\mathbf{E}_j \times \mathbf{E}_k}{\mathbf{E}_i \cdot (\mathbf{E}_j \times \mathbf{E}_k)}$ where $\mathbf{E}_i, \mathbf{E}_j, \mathbf{E}_k$ form a right handed orthonormal system. Hence, conclude that when the covariant or contravariant basis vector set is orthonormal then the covariant and contravariant components of a given tensor are identical.

Answer:
If $\mathbf{E}^i, \mathbf{E}^j, \mathbf{E}^k$ form a right handed orthonormal system then $\mathbf{E}^i \cdot (\mathbf{E}^j \times \mathbf{E}^k)$ (which is the volume of the parallelepiped formed by the vectors $\mathbf{E}^i, \mathbf{E}^j, \mathbf{E}^k$) equals 1 and hence we have:
$$\mathbf{E}_i = \mathbf{E}^j \times \mathbf{E}^k$$
However, since $\mathbf{E}^i, \mathbf{E}^j, \mathbf{E}^k$ form a right handed orthonormal system then we should also have:
$$\mathbf{E}^i = \mathbf{E}^j \times \mathbf{E}^k$$
On comparing the last two equations we conclude that $\mathbf{E}_i = \mathbf{E}^i$.
The second part of the question can be answered similarly by just exchanging the variance type of the involved vectors in the first part.
Now, since the covariant and contravariant basis vector sets are identical when the covariant or contravariant basis vector set is orthonormal then by the principle of invariance of tensors the covariant and contravariant components of a given tensor should also be identical because otherwise the tensor will vary depending on the employed basis set.

60. State the mathematical relations between the original and transformed (i.e. unbarred and barred) basis vector sets in their covariant and contravariant forms under admissible coordinate transformations.

    Answer: These relations are given by:
    $$\mathbf{E}_i = \frac{\partial \bar{u}^j}{\partial u^i} \bar{\mathbf{E}}_j \qquad\qquad \bar{\mathbf{E}}_i = \frac{\partial u^j}{\partial \bar{u}^i} \mathbf{E}_j$$
    $$\mathbf{E}^i = \frac{\partial u^i}{\partial \bar{u}^j} \bar{\mathbf{E}}^j \qquad\qquad \bar{\mathbf{E}}^i = \frac{\partial \bar{u}^i}{\partial u^j} \mathbf{E}^j$$
    where the indexed u and $\bar{u}$ represent the coordinates in the unbarred and barred systems, while the indexed $\mathbf{E}$ and $\bar{\mathbf{E}}$ are the basis vectors in the unbarred and barred systems.

61. Correct, if necessary, the following equations explaining all the symbols involved:
    $$\mathbf{E}_1 \cdot (\mathbf{E}_2 \times \mathbf{E}_3) = \frac{1}{\sqrt{g}} \qquad\qquad \mathbf{E}^1 \cdot (\mathbf{E}^2 \times \mathbf{E}^3) = \sqrt{g}$$

    Answer: The correct relations are:
    $$\mathbf{E}_1 \cdot (\mathbf{E}_2 \times \mathbf{E}_3) = \sqrt{g} \qquad\qquad \mathbf{E}^1 \cdot (\mathbf{E}^2 \times \mathbf{E}^3) = \frac{1}{\sqrt{g}}$$
    where $\mathbf{E}_1, \mathbf{E}_2, \mathbf{E}_3$ are the covariant basis vectors, $\mathbf{E}^1, \mathbf{E}^2, \mathbf{E}^3$ are the contravariant basis vectors and g is the determinant of the covariant metric tensor while the dot and cross represent the dot product and cross product operations of vectors.

62. Obtain the relation: $g = J^2$ from the relation: $\mathbf{J}^T\mathbf{J} = [g_{ij}]$ giving full explanation of each step.
    Answer: We start from the given relation (which is a definition):
    $$[g_{ij}] = \mathbf{J}^T\mathbf{J}$$
    where $[g_{ij}]$ is the matrix representing the covariant metric tensor and $\mathbf{J}$ and $\mathbf{J}^T$ are the Jacobian matrix and its transpose while the product on the right is a matrix product. By taking the determinant of both sides of this equation we obtain:
    $$g = J^T J$$
    where g, J^T, J are the determinants of the corresponding entities. The last relation is justified by the well known rule of linear algebra that the determinant of a product of matrices (i.e. $\mathbf{J}^T\mathbf{J}$) is equal to the product of the determinants of these matrices (i.e. $J^T J$). Now, according to another rule of linear algebra which states that the determinant of a matrix is equal to the determinant of its transpose we have $J^T = J$ and hence the last equation (i.e. $g = J^T J$) becomes:
    $$g = J^2$$
    as required.

63. State three consequences of having mutually orthogonal contravariant basis vectors at each point in the space justifying these consequences.
    Answer: The following are examples of these consequences (other consequences given in the text can also be quoted):
    (a) The covariant basis vectors should also be mutually orthogonal, i.e. $\mathbf{E}_i \cdot \mathbf{E}_j = 0$ when $i \neq j$. The reason is that the corresponding vectors of each basis set are in the same direction due to the fact that the tangent vector to the ith coordinate curve and the gradient vector of the ith coordinate surface at a given point in the space have the same direction and hence if the vectors of one set (i.e. covariant or contravariant) are mutually orthogonal then the vectors of the other set should also be mutually orthogonal.
    (b) The covariant and contravariant metric tensors are diagonal with non-vanishing diagonal elements, that is:
    $$g_{ij} = 0 \qquad g^{ij} = 0 \qquad (i \neq j)$$
    $$g_{ii} \neq 0 \qquad g^{ii} \neq 0 \qquad (\text{no sum on } i)$$
    The reason is that from the relations $g_{ij} = \mathbf{E}_i \cdot \mathbf{E}_j$ and $g^{ij} = \mathbf{E}^i \cdot \mathbf{E}^j$ we can see that the dot product (and hence the element of the metric tensor) is zero when the indices are different due to the mutual orthogonality of the basis vectors. Moreover, the dot product (and hence the element of the metric tensor) should be non-zero when the indices are identical because the basis vectors cannot vanish at the regular points of the space since the tangent to the coordinate curve and the gradient to the coordinate surface do exist and they cannot be zero.

(c) The diagonal elements of the covariant metric tensor and the corresponding elements of the contravariant metric tensor are reciprocals of each other, that is:

$$g^{ii} = \frac{1}{g_{ii}} \qquad \text{(no summation on } i\text{)}$$

The reason is that since the covariant and contravariant metric tensors are inverses of each other and they are diagonal with non-vanishing diagonal elements (as established in the previous point) then their corresponding diagonal elements should be reciprocal of each other (as proved in linear algebra).

64. Discuss the relationship between the concepts of space, coordinate system and metric tensor.

    Answer: The required answer to this question should be no more than a short summary of section 2.7 (Relationship between Space, Coordinates and Metric) of the book. The essence of this summary is the need of any abstract space for a coordinate system to identify its points and describe its properties and this will lead to the emergence of the metric tensor as a mathematical entity that identifies and characterizes the geometric properties of the space locally and globally.

Chapter 3
Tensors

1. Define "covariant" and "contravariant" tensors from the perspective of their notation and their transformation rules.
   Answer: Covariant tensors are notated with subscript indices while contravariant tensors are notated with superscript indices. A covariant tensor of rank-m is transformed according to the following rule:

$$\bar{A}_{ij\cdots m} = \frac{\partial u^p}{\partial \bar{u}^i} \frac{\partial u^q}{\partial \bar{u}^j} \cdots \frac{\partial u^r}{\partial \bar{u}^m} A_{pq\cdots r}$$

   while a contravariant tensor of rank-m is transformed according to the following rule:

$$\bar{B}^{ij\cdots m} = \frac{\partial \bar{u}^i}{\partial u^p} \frac{\partial \bar{u}^j}{\partial u^q} \cdots \frac{\partial \bar{u}^m}{\partial u^r} B^{pq\cdots r}$$

2. Write the transformation relations for covariant and contravariant vectors and for covariant, contravariant and mixed rank-2 tensors between different coordinate systems.
   Answer: The transformation relations for covariant and contravariant vectors are:

$$\bar{A}_i = \frac{\partial u^j}{\partial \bar{u}^i} A_j \qquad\qquad \bar{B}^i = \frac{\partial \bar{u}^i}{\partial u^j} B^j$$

   The transformation relations for covariant, contravariant and mixed rank-2 tensors are:

$$\bar{A}_{ij} = \frac{\partial u^p}{\partial \bar{u}^i} \frac{\partial u^q}{\partial \bar{u}^j} A_{pq} \qquad \bar{B}^{ij} = \frac{\partial \bar{u}^i}{\partial u^p} \frac{\partial \bar{u}^j}{\partial u^q} B^{pq} \qquad \bar{C}_i{}^j = \frac{\partial u^p}{\partial \bar{u}^i} \frac{\partial \bar{u}^j}{\partial u^q} C_p{}^q$$

3. State the practical rules for writing the transformation relations of tensors between different coordinate systems.
   Answer: The practical rules can be summarized as follows where we transform from unbarred system to barred system (using $A^i{}_j{}^k$ as an example of a tensor that we transform):
   • Write the symbol of the transformed tensor on the left hand side of the transformation equation and the symbol of the original tensor on the right hand side:

$$\bar{A} = A$$

   • Index the original tensor with its original indices and index the transformed tensor with different indices noting that its indicial structure should be similar to the indicial structure of the original tensor:

$$\bar{A}^l{}_m{}^n = A^i{}_j{}^k$$

3   TENSORS

- Insert a number of partial differential operators on the right hand side equal to the number of free indices:

$$\bar{A}^l{}_m{}^n = \frac{\partial u}{\partial u}\frac{\partial u}{\partial u}\frac{\partial u}{\partial u} A^i{}_j{}^k$$

- Index the coordinates of the transformed tensor in the numerator or denominator in these operators according to the order of the indices in the tensor where these indices are in the same position (upper or lower) as their position in the tensor:

$$\bar{A}^l{}_m{}^n = \frac{\partial u^l}{\partial u}\frac{\partial u}{\partial u^m}\frac{\partial u^n}{\partial u} A^i{}_j{}^k$$

- Because the transformed tensor is barred then its coordinates should also be barred:

$$\bar{A}^l{}_m{}^n = \frac{\partial \bar{u}^l}{\partial u}\frac{\partial u}{\partial \bar{u}^m}\frac{\partial \bar{u}^n}{\partial u} A^i{}_j{}^k$$

- Index the coordinates of the original tensor in the numerator or denominator in these operators according to the order of the indices in the tensor where these indices are in the opposite position (upper or lower) to their position in the tensor:

$$\bar{A}^l{}_m{}^n = \frac{\partial \bar{u}^l}{\partial u^i}\frac{\partial u^j}{\partial \bar{u}^m}\frac{\partial \bar{u}^n}{\partial u^k} A^i{}_j{}^k$$

4. What are the raising and lowering operators and how they provide the link between the covariant and contravariant types?
Answer: The raising operator is the contravariant metric tensor while the lowering operator is the covariant metric tensor. The raising operator can change covariant indices to contravariant indices while the lowering operator can change contravariant indices to covariant indices. Since these operators can change a tensor from one variance type to another, they provide a link between the different variance types of the tensor and hence they facilitate the transformation between different basis sets of a given coordinate system.

5. $\mathbf{A}$ is a tensor of type (m, n) and $\mathbf{B}$ is a tensor of type (p, q, w). What this means? Write these tensors in their indicial form.
Answer: It means that $\mathbf{A}$ is a tensor with m contravariant indices and n covariant indices, and $\mathbf{B}$ is a tensor with p contravariant indices, q covariant indices and weight w. In indicial form, these tensors should be written as $A^{i_1 \cdots i_m}_{j_1 \cdots j_n}$ and $B^{i_1 \cdots i_p}_{j_1 \cdots j_q}$.

6. Write the following equations in full tensor notation and explain their significance:

$$\mathbf{E}_i = \frac{\partial \mathbf{r}}{\partial u^i} \qquad\qquad \mathbf{E}^i = \nabla u^i$$

Answer:

$$\mathbf{E}_i = \frac{\partial x^j}{\partial u^i}\mathbf{e}_j \qquad\qquad \mathbf{E}^i = \frac{\partial u^i}{\partial x^j}\mathbf{e}_j$$

where $\mathbf{E}_i$ and $\mathbf{E}^i$ are covariant and contravariant general basis vectors, x^j are Cartesian coordinates, u^i are general coordinates and $\mathbf{e}_j$ are Cartesian basis vectors.

The significance of these equations is that the covariant basis vectors are tangents to the coordinate curves while the contravariant basis vectors are gradients to the coordinate surfaces.

7. Write the orthonormalized form of the covariant basis vectors in a 2D general coordinate system. Verify that these vectors are actually orthonormal.
Answer: They are:

$$\underline{\mathbf{E}}_1 = \frac{\mathbf{E}_1}{|\mathbf{E}_1|} = \frac{\mathbf{E}_1}{\sqrt{g_{11}}} \qquad \underline{\mathbf{E}}_2 = \frac{g_{11}\mathbf{E}_2 - g_{12}\mathbf{E}_1}{\sqrt{g_{11}g}}$$

where $\underline{\mathbf{E}}_1$ and $\underline{\mathbf{E}}_2$ are orthonormalized covariant basis vectors, $\mathbf{E}_1$ and $\mathbf{E}_2$ are general covariant basis vectors, the indexed g are coefficients of the covariant metric tensor and g is its determinant.

Verification:

$$\begin{aligned}
\underline{\mathbf{E}}_1 \cdot \underline{\mathbf{E}}_1 &= \frac{\mathbf{E}_1}{|\mathbf{E}_1|} \cdot \frac{\mathbf{E}_1}{|\mathbf{E}_1|} \\
&= \frac{\mathbf{E}_1 \cdot \mathbf{E}_1}{|\mathbf{E}_1|^2} \\
&= \frac{|\mathbf{E}_1|^2}{|\mathbf{E}_1|^2} \\
&= 1
\end{aligned}$$

$$\begin{aligned}
\underline{\mathbf{E}}_2 \cdot \underline{\mathbf{E}}_2 &= \frac{g_{11}\mathbf{E}_2 - g_{12}\mathbf{E}_1}{\sqrt{g_{11}g}} \cdot \frac{g_{11}\mathbf{E}_2 - g_{12}\mathbf{E}_1}{\sqrt{g_{11}g}} \\
&= \frac{(g_{11}\mathbf{E}_2 - g_{12}\mathbf{E}_1) \cdot (g_{11}\mathbf{E}_2 - g_{12}\mathbf{E}_1)}{g_{11}g} \\
&= \frac{g_{11}g_{11}\mathbf{E}_2 \cdot \mathbf{E}_2 - g_{11}g_{12}\mathbf{E}_2 \cdot \mathbf{E}_1 - g_{12}g_{11}\mathbf{E}_1 \cdot \mathbf{E}_2 + g_{12}g_{12}\mathbf{E}_1 \cdot \mathbf{E}_1}{g_{11}g} \\
&= \frac{g_{11}g_{11}g_{22} - g_{11}g_{12}g_{21} - g_{12}g_{11}g_{12} + g_{12}g_{12}g_{11}}{g_{11}g} \\
&= \frac{g_{11}g_{11}g_{22} - g_{11}g_{12}g_{21}}{g_{11}g} \\
&= \frac{g_{11}(g_{11}g_{22} - g_{12}g_{21})}{g_{11}g} \\
&= \frac{g_{11}g}{g_{11}g} \\
&= 1
\end{aligned}$$

$$\begin{aligned}
\underline{\mathbf{E}}_1 \cdot \underline{\mathbf{E}}_2 &= \frac{\mathbf{E}_1}{\sqrt{g_{11}}} \cdot \frac{g_{11}\mathbf{E}_2 - g_{12}\mathbf{E}_1}{\sqrt{g_{11}g}} \\
&= \frac{\mathbf{E}_1 \cdot (g_{11}\mathbf{E}_2 - g_{12}\mathbf{E}_1)}{g_{11}\sqrt{g}}
\end{aligned}$$

3 TENSORS

$$= \frac{g_{11}\mathbf{E}_1 \cdot \mathbf{E}_2 - g_{12}\mathbf{E}_1 \cdot \mathbf{E}_1}{g_{11}\sqrt{g}}$$

$$= \frac{g_{11}g_{12} - g_{12}g_{11}}{g_{11}\sqrt{g}}$$

$$= 0$$

Hence, the vectors $\underline{\mathbf{E}}_1$ and $\underline{\mathbf{E}}_2$ are orthonormal (i.e. they are orthogonal to each other according to the third dot product and of unit length according to the first and second dot products).

8. Why the following relations are labeled as the reciprocity relations?

$$\mathbf{E}_i \cdot \mathbf{E}^j = \delta_i^j \qquad\qquad \mathbf{E}^i \cdot \mathbf{E}_j = \delta^i_j$$

Answer: Because these relations express the fact that the covariant and contravariant basis vectors are reciprocal systems. It is noteworthy that two sets of vectors of the same number of elements (e.g. $\mathbf{V}_1, \mathbf{V}_2, \cdots \mathbf{V}_n$ and $\mathbf{W}_1, \mathbf{W}_2, \cdots \mathbf{W}_n$)[7] are described as reciprocal to each other if they satisfy the following relations:

$$\mathbf{V}_i \cdot \mathbf{W}_j = 1 \qquad (i = j)$$
$$\mathbf{V}_i \cdot \mathbf{W}_j = 0 \qquad (i \neq j)$$

9. The components of the tensors $\mathbf{A}$, $\mathbf{B}$ and $\mathbf{C}$ are given by: $A_{ik}^{\cdot\cdot j}$, $B^{jn}_{\cdot\cdot mq}$ and $C_{k\cdot i}^{\cdot l\cdot}$. Write these tensors in their full notation that includes their basis tensors.
Answer:

$$\mathbf{A} = A_{ik}^{\cdot\cdot j}\mathbf{E}^i\mathbf{E}^k\mathbf{E}_j \qquad \mathbf{B} = B^{jn}_{\cdot\cdot mq}\mathbf{E}_j\mathbf{E}_n\mathbf{E}^m\mathbf{E}^q \qquad \mathbf{C} = C_{k\cdot i}^{\cdot l\cdot}\mathbf{E}^k\mathbf{E}_l\mathbf{E}^i$$

10. $\mathbf{A}$, $\mathbf{B}$ and $\mathbf{C}$ are tensors of rank-2, rank-3 and rank-4 respectively in a given coordinate system. Write the components of these tensors with respect to the following basis tensors: , $\mathbf{E}^i\mathbf{E}^n$, $\mathbf{E}_i\mathbf{E}_k\mathbf{E}_m$ and $\mathbf{E}_j\mathbf{E}^i\mathbf{E}_k\mathbf{E}^n$.
Answer:

$$\mathbf{A} = A_{in}\mathbf{E}^i\mathbf{E}^n \qquad \mathbf{B} = B^{ikm}\mathbf{E}_i\mathbf{E}_k\mathbf{E}_m \qquad \mathbf{C} = C^{j\cdot k\cdot}_{\cdot i\cdot n}\mathbf{E}_j\mathbf{E}^i\mathbf{E}_k\mathbf{E}^n$$

11. What "dyad" means? Write all the nine unit dyads associated with the double directions of rank-2 tensors in a 3D space with a rectangular Cartesian coordinate system (i.e. $\mathbf{e}_1\mathbf{e}_1 \cdots \mathbf{e}_3\mathbf{e}_3$).
Answer: Dyad is a rank-2 tensor obtained by the direct multiplication of two vectors. The nine unit dyads are:

$$\begin{array}{ccc}
\mathbf{e}_1\mathbf{e}_1 & \mathbf{e}_1\mathbf{e}_2 & \mathbf{e}_1\mathbf{e}_3 \\
\mathbf{e}_2\mathbf{e}_1 & \mathbf{e}_2\mathbf{e}_2 & \mathbf{e}_2\mathbf{e}_3 \\
\mathbf{e}_3\mathbf{e}_1 & \mathbf{e}_3\mathbf{e}_2 & \mathbf{e}_3\mathbf{e}_3
\end{array}$$

12. Make a simple sketch of the nine dyads of exercise 11.
Answer: The sketch should look like Figure 4.

---

[7] The subscript indices are meant to label the vectors with no significance about their variance type.

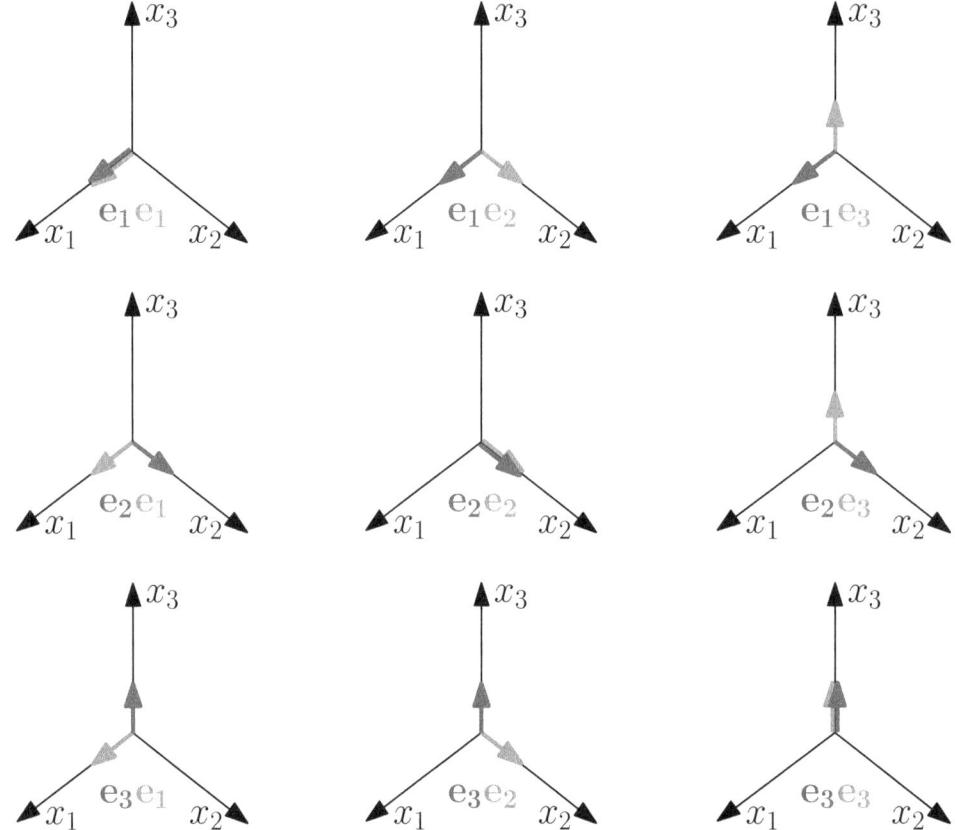

Figure 4: The nine unit dyads associated with the double directions of rank-2 tensors in a 3D space with a rectangular Cartesian coordinate system.

13. Compare true and pseudo vectors making a clear distinction between the two with a simple illustrating plot. Generalize this to tensors of any rank.
    Answer: True vectors transform invariantly under coordinate transformations and hence they keep their direction, while pseudo vectors do not transform invariantly under improper orthogonal transformations which involve inversion of coordinate axes through the origin of coordinates with a change of system handedness since they acquire a minus sign under such transformations and hence they reverse their direction. The plot should look like Figure 5 where we see a true vector **v** that keeps its direction in the space following a reflection of the coordinate system through the origin of coordinates and a pseudo vector **p** that reverses its direction following this operation. To generalize these properties to tensors of any rank, we simply replace the reversal of the single direction in the above definition of true and pseudo vectors with the change of multi directions that associate tensors of higher ranks.
14. Justify the following statement: "The terms of consistent tensor expressions and equations should be uniform in their true and pseudo type".
    Answer: Let assume that we have tensor expressions/equations with mixed terms (i.e. some true and some pseudo) and we transform these expressions/equations by

3   TENSORS   39

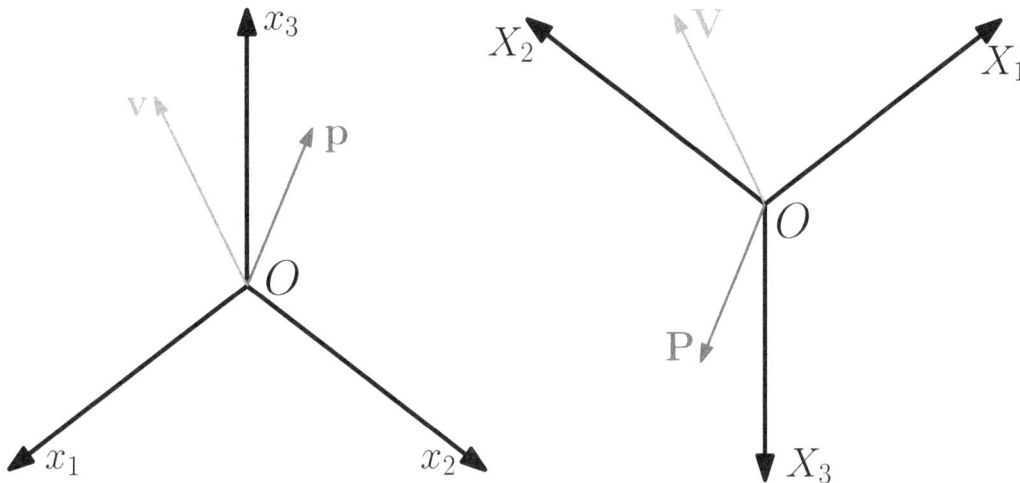

Figure 5: The behavior of a true vector (**v** and **V**) and a pseudo vector (**p** and **P**) following a reflection of the coordinate system in the origin of coordinates.

an improper orthogonal transformation, then the true terms will transform invariantly while the pseudo terms will not, and this does not make sense because these expressions/equations will behave neither as true tensors nor as pseudo tensors and hence they have indeterminate state.

15. What is the curl of a pseudo vector from the perspective of true/pseudo qualification?
    Answer: It should be a true vector.

16. Define absolute and relative tensors stating any necessary mathematical relations.
    Answer: Absolute tensors are those tensors that have no Jacobian factor in their transformation equation, while relative tensors have such a factor. In mathematical terms, the transformation equation of these two types of tensor is:
    $$\bar{A}^{ij\ldots k}{}_{lm\ldots n} = \left|\frac{\partial x}{\partial \bar{x}}\right|^w \frac{\partial \bar{x}^i}{\partial x^a} \frac{\partial \bar{x}^j}{\partial x^b} \cdots \frac{\partial \bar{x}^k}{\partial x^c} \frac{\partial x^d}{\partial \bar{x}^l} \frac{\partial x^e}{\partial \bar{x}^m} \cdots \frac{\partial x^f}{\partial \bar{x}^n} A^{ab\ldots c}{}_{de\ldots f}$$
    where $w = 0$ for absolute tensor and $w \neq 0$ for relative tensor.

17. What is the weight of the product of **A** and **B** where **A** is a tensor of type $(1, 2, 2)$ and **B** is a tensor of type $(0, 3, -1)$?
    Answer: The weight of **A** is 2 and the weight of **B** is -1 and hence the weight of their product is the sum of their weights, i.e. $w = 2 - 1 = 1$.

18. Show that the determinant of a rank-2 absolute tensor **A** is a relative scalar and find the weight in the case of **A** being covariant and in the case of **A** being contravariant.
    Answer: The transformation equation of a rank-2 absolute covariant tensor is given by:
    $$\bar{A}_{ij} = \frac{\partial u^p}{\partial \bar{u}^i} \frac{\partial u^q}{\partial \bar{u}^j} A_{pq}$$

On taking the determinant of both sides we obtain:[8]

$$\left|\bar{A}_{ij}\right| = \left|\frac{\partial u^p}{\partial \bar{u}^i}\right|\left|\frac{\partial u^q}{\partial \bar{u}^j}\right||A_{pq}| = J^2\,|A_{pq}|$$

and hence the determinant of a rank-2 absolute covariant tensor **A** is a relative scalar of weight 2.[9]

Similarly, the transformation equation of a rank-2 absolute contravariant tensor is given by:

$$\bar{A}^{ij} = \frac{\partial \bar{u}^i}{\partial u^p}\frac{\partial \bar{u}^j}{\partial u^q}A^{pq}$$

On taking the determinant of both sides we obtain:

$$\left|\bar{A}^{ij}\right| = \left|\frac{\partial \bar{u}^i}{\partial u^p}\right|\left|\frac{\partial \bar{u}^j}{\partial u^q}\right||A^{pq}| = \left(J^{-1}\right)^2|A^{pq}| = J^{-2}\,|A^{pq}|$$

and hence the determinant of a rank-2 absolute contravariant tensor **A** is a relative scalar of weight -2.

19. Why the tensor terms of tensor expressions and equalities should have the same weight?
    Answer: The weight is a characteristic property of the tensor and its transformation rules and hence for consistency and homogeneity the terms of tensor expressions and equalities should have the same weight. We can repeat our previous argument about the necessity of consistency of terms with regard to their true and pseudo nature and hence if we transform a tensor expression or equality whose terms have different weights then we will not have a consistent transformation rule because different terms require different weights.

20. What "isotropic" and "anisotropic" tensor mean?
    Answer: Isotropic tensors are those tensors whose components do not change under proper rotational transformations, while anisotropic tensors are those tensors whose components do change under such transformations.

21. Give an example of an isotropic rank-2 tensor and another example of an anisotropic rank-3 tensor.
    Answer: The Kronecker delta is an example of an isotropic rank-2 tensor, while the piezoelectric moduli tensor is an example of an anisotropic rank-3 tensor.

22. What is the significance of the fact that the zero tensor of all ranks and all dimensions is isotropic with regard to the invariance of tensors under coordinate transformations?
    Answer: The significance is that if the components of a tensor vanish in a particular coordinate system then they will vanish in all properly and improperly rotated coordinate systems.[10] As a result, if the components of two tensors are equal in a particular

---

[8] The symbol | | stands for the determinant of the enclosed matrix.

[9] We are referring here to the equation of Exercise 16 where a scalar f should transform (according to that equation) as:

$$\bar{f} = J^w f$$

and hence $w = 0$ for absolute scalar and $w \neq 0$ for relative scalar.

[10] For improper rotation, this is more general than being isotropic.

coordinate system then they should be equal in all other systems since the tensor of their difference is a zero tensor and hence it is invariant. This leads to the conclusion that identities and equalities of tensors are invariant under coordinate transformations.

23. Define "symmetric" and "anti-symmetric" tensor. Why scalars and vectors are not qualified to be symmetric or anti-symmetric?

    Answer: Symmetric tensors are those tensors whose component values do not change under certain exchange of indices, while anti-symmetric tensors are those tensors whose component values reverse their sign (i.e. identical magnitude with opposite sign) under certain exchange of indices. Accordingly, if A_{ij} is a symmetric tensor and B_{ij} is an anti-symmetric tensor, then their components will satisfy the following relations:

$$A_{ji} = +A_{ij}$$
$$B_{ji} = -B_{ij}$$

    Because symmetry and anti-symmetry require at least two indices to have an exchange, then scalars with no index and vectors with just one index cannot be symmetric or anti-symmetric since no exchange of indices can be imagined.

24. Write the symmetric and anti-symmetric parts of the tensor A^{ij}.

    Answer:

$$A^{(ij)} = \frac{1}{2}\left(A^{ij} + A^{ji}\right) \qquad \text{(Symmetric part)}$$
$$A^{[ij]} = \frac{1}{2}\left(A^{ij} - A^{ji}\right) \qquad \text{(Anti-symmetric part)}$$

25. Write the symmetrization and anti-symmetrization formulae for a rank-n tensor $A^{i_1 \ldots i_n}$.

    Answer:
    Symmetrization formula:

$$A^{(i_1 \ldots i_n)} = \frac{1}{n!}\left(\sum \text{even permutations of } i\text{'s} + \sum \text{odd permutations of } i\text{'s}\right)$$

    Anti-symmetrization formula:

$$A^{[i_1 \ldots i_n]} = \frac{1}{n!}\left(\sum \text{even permutations of } i\text{'s} - \sum \text{odd permutations of } i\text{'s}\right)$$

26. Symmetrize and anti- symmetrize the tensor A^{ijkl} with respect to its second and fourth indices.

    Answer:

$$A^{i(j)k(l)} = \frac{1}{2}\left(A^{ijkl} + A^{ilkj}\right)$$
$$A^{i[j]k[l]} = \frac{1}{2}\left(A^{ijkl} - A^{ilkj}\right)$$

27. Write the two mathematical conditions for a rank-n tensor $A^{i_1...i_n}$ to be totally symmetric and totally anti-symmetric.
    Answer: The conditions are:

    $$A^{i_1...i_n} = A^{(i_1...i_n)} \qquad \text{(Totally symmetric)}$$
    $$A^{i_1...i_n} = A^{[i_1...i_n]} \qquad \text{(Totally anti-symmetric)}$$

    Similarly:

    $$A^{[i_1...i_n]} = 0 \qquad \text{(Totally symmetric)}$$
    $$A^{(i_1...i_n)} = 0 \qquad \text{(Totally anti-symmetric)}$$

28. The tensor A_{ijk} is totally symmetric. How many distinct components it has in a 3D space?
    Answer: In a 3D space, A_{ijk} has $3^3 = 27$ components which correspond to all the possible permutations (including the repetitive ones) of the numbers 1, 2, 3. However, due to the total symmetry the order of the numbers is irrelevant and hence what we need is to extract all the possible combinations (including the repetitive ones) of the numbers 1, 2, 3, that is:
    Combinations with only one distinct number: 111, 222, 333.
    Combinations with only two distinct numbers: 112, 113, 122, 223, 133, 233.
    Combinations with three distinct numbers: 123.
    Accordingly, A_{ijk} should have 10 distinct components.[11]

29. The tensor B_{ijk} is totally anti-symmetric. How many identically vanishing components it has in a 3D space? How many distinct non-identically vanishing components it has in a 3D space?
    Answer: In a 3D space, B_{ijk} has $3^3 = 27$ components which correspond to all the possible permutations (including the repetitive ones) of the numbers 1, 2, 3. Now, since it is totally anti-symmetric then any permutation with two identical numbers should be zero. So, all permutations except the permutations of the combination 123 should be zero. Because the combination 123 has six permutations then we should have $27 - 6 = 21$ identically vanishing components and 6 non-identically vanishing components. These 6 non-identically vanishing components correspond to 3 even permutations of 123 and 3 odd permutations of 123. The components of the 3 even permutations of 123 should be identical and the components of the 3 odd permutations of 123 should be identical. Hence, B_{ijk} should have two distinct non-identically vanishing components where these two differ only in sign. However, since "distinct" in such contexts means "independent" then we should have only one distinct non-identically vanishing component because the components corresponding to the even and odd permutations differ only in sign.
    All these results can be obtained from inspecting the rank-3 permutation tensor.

30. Give numeric or symbolic examples of a rank-2 symmetric tensor and a rank-2 skew-symmetric tensor in a 4D space. Count the number of independent non-identically

---

[11] In this context, "distinct" means "independent".

3 TENSORS   43

vanishing components in each case.
Answer:
Symmetric:
$$\begin{matrix} a & e & f & h \\ e & b & g & i \\ f & g & c & j \\ h & i & j & d \end{matrix}$$

The number of independent non-identically vanishing components is $\frac{n(n+1)}{2} = 10$.
Skew-symmetric:
$$\begin{matrix} 0 & -1 & -2 & -4 \\ 1 & 0 & -3 & -5 \\ 2 & 3 & 0 & -6 \\ 4 & 5 & 6 & 0 \end{matrix}$$

The number of independent non-identically vanishing components is $\frac{n(n-1)}{2} = 6$.

31. Write the formula for the number of independent components of a rank-2 symmetric tensor, and the formula for the number of independent non-zero components of a rank-2 anti-symmetric tensor in nD space.
Answer: The formulae are:
$$N_s = \frac{n(n+1)}{2}$$
$$N_a = \frac{n(n-1)}{2}$$

where N_s is the number of independent components of a rank-2 symmetric tensor and N_a is the number of independent non-zero (i.e. non-identically vanishing) components of a rank-2 anti-symmetric tensor.

32. Explain why the entries corresponding to identical anti-symmetric indices should vanish identically.
Answer: Because an exchange of two identical indices, which identifies the same entry, should change the sign of the entry due to anti-symmetry and hence the entry should be equal to its negative and this can be true only if the entry is identically zero.

33. Why the indices whose exchange defines the symmetry and anti-symmetry relations should be of the same variance type?
Answer: Because the covariant and contravariant indices correspond to different basis sets (i.e. contravariant and covariant respectively) and hence even if the values of the components satisfy the property of symmetry or anti-symmetry it does not lead to the symmetry or anti-symmetry of the tensor due to the involvement of different basis sets between which no symmetry or anti-symmetry can be defined sensibly. For example, let have a rank-2 mixed tensor **A** whose component values satisfy the relation $A^i{}_j = A^j{}_i$. However, since the tensor components $A^i{}_j$ mean $A^i{}_j \mathbf{E}_i \mathbf{E}^j$ while $A^j{}_i$ mean $A^j{}_i \mathbf{E}_j \mathbf{E}^i$ then no symmetry or anti-symmetry can be defined for the tensor because no symmetry or anti-symmetry can be defined sensibly for $\mathbf{E}_i \mathbf{E}^j$ and $\mathbf{E}_j \mathbf{E}^i$ due to the fact they involve

3 TENSORS 44

vectors of opposite variance type. This is unlike the situation with a rank-2 covariant tensor **B** for example because even though the component B_{ij} means $B_{ij}\mathbf{E}^i\mathbf{E}^j$ while the component B_{ji} means $B_{ji}\mathbf{E}^j\mathbf{E}^i$ symmetry or anti-symmetry can be defined for the tensor since symmetry or anti-symmetry can be defined for the unique basis set, i.e. $\mathbf{E}^i\mathbf{E}^j$ and $\mathbf{E}^j\mathbf{E}^i$ can be seen as a symmetric basis tensor since both basis vectors belong to the same basis set.

To put it in more simple terms, when the two indices are of the same variance type then exchanging the indices is achieved by just exchanging their values and hence we have $B_{ij}\mathbf{E}^i\mathbf{E}^j$ and $B_{ji}\mathbf{E}^j\mathbf{E}^i$ are symmetric or anti-symmetric in a sensible way, but when the two indices are of different variance type then exchanging the indices is not sufficient to achieve symmetry or anti-symmetry since the order of the two indices cannot be changed (due to their different variance type) because when we exchange the indices of $A^i{}_j\mathbf{E}_i\mathbf{E}^j$ we get $A^j{}_i\mathbf{E}_j\mathbf{E}^i$ (rather than $A_i{}^j\mathbf{E}^i\mathbf{E}_j$ or $A_j{}^i\mathbf{E}^j\mathbf{E}_i$) and hence sensible symmetry or anti-symmetry cannot be achieved.

34. Discuss the significance of the fact that the symmetry and anti-symmetry characteristic of a tensor is invariant under coordinate transformations and link this to the invariance of the zero tensor.
    Answer: The significance is that by knowing that a tensor is symmetric or anti-symmetric or neither in one system we know that it is symmetric or anti-symmetric or neither in all other systems with no extra effort to establish its status from this perspective. The link between this invariance property and the invariance of the zero tensor is that the difference between a tensor and its symmetric associate is the zero tensor while the sum of a tensor and its anti-symmetric associate is the zero tensor and because the zero tensor is invariant across all coordinate systems then the difference or sum should also be invariant across all coordinate systems and this leads to the conclusion that the symmetric or anti-symmetric property of a tensor should be invariant across all coordinate systems.[12]

35. Verify the relation $A_{ij}B^{ij} = 0$, where A_{ij} is a symmetric tensor and B^{ij} is an anti-symmetric tensor, by writing the sum in full assuming a 3D space.
    Answer: We write the sum in a matrix-like form, that is:

$$\begin{aligned} A_{ij}B^{ij} &= A_{11}B^{11} + A_{12}B^{12} + A_{13}B^{13} + \\ &\quad A_{21}B^{21} + A_{22}B^{22} + A_{23}B^{23} + \\ &\quad A_{31}B^{31} + A_{32}B^{32} + A_{33}B^{33} \end{aligned}$$

Now, because A_{ij} is symmetric and B^{ij} is anti-symmetric then the corresponding terms across the main diagonal (e.g. $A_{13}B^{13}$ versus $A_{31}B^{31}$ with $A_{13} = A_{31}$ and $B^{13} = -B^{31}$) should be equal in magnitude and opposite in sign, that is:

$$\begin{aligned} A_{ij}B^{ij} &= A_{11}B^{11} - A_{21}B^{21} - A_{31}B^{31} + \\ &\quad A_{21}B^{21} + A_{22}B^{22} - A_{32}B^{32} + \end{aligned}$$

---

[12] In fact, this is not a rigorous proof but rather a demonstration since certain restrictions and extensions (e.g. with regard to partial symmetry or anti-symmetry) are needed to obtain the required result.

$$A_{31}B^{31} + A_{32}B^{32} + A_{33}B^{33}$$

that is:
$$A_{ij}B^{ij} = A_{11}B^{11} + A_{22}B^{22} + A_{33}B^{33}$$

Finally, because B^{ij} is anti-symmetric then $B^{11} = B^{22} = B^{33} = 0$ and hence we have:
$$A_{ij}B^{ij} = 0 + 0 + 0 = 0$$

as required.

36. Classify the common tensor operations with respect to the number of tensors involved in these operations.
    Answer: Addition and subtraction, multiplication of tensor by scalar, tensor multiplication and inner product involve two tensors (considering scalar as rank-0 tensor). Permutation involves one tensor. Contraction can involve one tensor or two tensors.

37. Which of the following operations are commutative, associative or distributive when these properties apply: algebraic addition, algebraic subtraction, multiplication by a scalar, outer multiplication, and inner multiplication?
    Answer:
    Algebraic addition is commutative and associative.
    Algebraic subtraction is neither commutative nor associative.
    Multiplication by a scalar is commutative, associative and distributive over algebraic addition and algebraic subtraction.
    Outer multiplication is not commutative but it is distributive over algebraic addition and algebraic subtraction.
    Inner multiplication is generally like outer multiplication in this respect (with some exceptions such as the commutativity of multiplication of two vectors).

38. For question 37, write all the required mathematical relations that describe those properties.
    Answer:
    Algebraic addition:
    $$\mathbf{A} + \mathbf{B} = \mathbf{B} + \mathbf{A}$$
    $$(\mathbf{A} + \mathbf{B}) + \mathbf{C} = \mathbf{A} + (\mathbf{B} + \mathbf{C})$$

    Algebraic subtraction:
    $$\mathbf{A} - \mathbf{B} \neq \mathbf{B} - \mathbf{A}$$
    $$(\mathbf{A} - \mathbf{B}) - \mathbf{C} \neq \mathbf{A} - (\mathbf{B} - \mathbf{C})$$

    Multiplication by a scalar:
    $$a\mathbf{A} = \mathbf{A}a$$
    $$a(b\mathbf{A}) = (ab)\mathbf{A}$$
    $$a(\mathbf{A} \pm \mathbf{B}) = a\mathbf{A} \pm a\mathbf{B}$$

Outer multiplication:

$$\mathbf{AB} \neq \mathbf{BA}$$
$$\mathbf{A}\left(\mathbf{B} \pm \mathbf{C}\right) = \mathbf{AB} \pm \mathbf{AC}$$

Inner multiplication:

$$\mathbf{A} \cdot \mathbf{B} \neq \mathbf{B} \cdot \mathbf{A}$$
$$\mathbf{A} \cdot \left(\mathbf{B} \pm \mathbf{C}\right) = \mathbf{A} \cdot \mathbf{B} \pm \mathbf{A} \cdot \mathbf{C}$$

39. The tensors involved in tensor addition, subtraction or equality should be compatible in their types. Give all the details about these "types".
    Answer: The involved tensors should have the same rank, the same space dimension, the same indicial structure (e.g. same set of free indices, same variance type of each index and same order of indices), and the same basis vector set (i.e. the set, whether covariant or contravariant, belongs to the same coordinate system). They should also be of the same true/pseudo type and have the same weight w whether $w = 0$ (absolute) or $w \neq 0$ (relative).[13]

40. What is the meaning of multiplying a tensor by a scalar in terms of the components of the tensor?
    Answer: Multiplying a tensor by a scalar means multiplying each component of the tensor by that scalar and hence it is a simple uniform scaling of the tensor (assuming the scalar is not zero).

41. A tensor of type (m_1, n_1, w_1) is multiplied by another tensor of type (m_2, n_2, w_2). What is the type, the rank and the weight of the product?
    Answer: The type is $(m_1 + m_2, n_1 + n_2, w_1 + w_2)$, the rank is $m_1 + m_2 + n_1 + n_2$ and the weight is $w_1 + w_2$.

42. We have two tensors: $\mathbf{A} = A_{ij}\mathbf{E}^i\mathbf{E}^j$ and $\mathbf{B} = B^{kl}\mathbf{E}_k\mathbf{E}_l$. We also have $\mathbf{C} = \mathbf{AB}$ and $\mathbf{D} = \mathbf{BA}$. Use the properties of tensor operations to obtain the full expression of $\mathbf{C}$ and $\mathbf{D}$ in terms of their components and basis tensors (i.e. $\mathbf{C} = \mathbf{AB} = \cdots$ etc.).
    Answer:

$$\mathbf{C} = \mathbf{AB} = A_{ij}\mathbf{E}^i\mathbf{E}^j B^{kl}\mathbf{E}_k\mathbf{E}_l = A_{ij}B^{kl}\mathbf{E}^i\mathbf{E}^j\mathbf{E}_k\mathbf{E}_l$$
$$\mathbf{D} = \mathbf{BA} = B^{kl}\mathbf{E}_k\mathbf{E}_l A_{ij}\mathbf{E}^i\mathbf{E}^j = B^{kl}A_{ij}\mathbf{E}_k\mathbf{E}_l\mathbf{E}^i\mathbf{E}^j$$

43. Explain why tensor multiplication, unlike ordinary multiplication of scalars, is not commutative considering the basis tensors to which the tensors are referred.
    Answer: Referring to the previous exercise, we see that the order of the basis vectors depends on the order of the multiplied tensors and since the order of the basis vectors is significant in determining the basis tensor (e.g. $\mathbf{E}^j\mathbf{E}_k \neq \mathbf{E}_k\mathbf{E}^j$) and cannot be changed or reversed arbitrarily, then tensor multiplication is not commutative. This is unlike ordinary multiplication of scalars since scalars are not associated with basis vectors.

---

[13] We note that some of these conditions are overlapping; however we prefer to put it in this way for clarity.

44. The direct product of vectors **a** and **b** is **ab**. Edit the following equation by adding a simple notation to make it correct without changing the order: **ab** = **ba**.
   Answer: We simply convert this to an equation of inner produce by adding a dot between the vectors, that is:
   $$\mathbf{a} \cdot \mathbf{b} = \mathbf{b} \cdot \mathbf{a}$$
   The justification is that while the direct product of vectors is not commutative, the inner product of vectors is commutative.

45. What is the difference in notation between matrix multiplication and tensor multiplication of two tensors, **A** and **B**, when we write **AB**?
   Answer: In matrix notation the matrix multiplication **AB** represents an inner product operation, while in tensor notation the tensor multiplication **AB** represents an outer product operation.

46. Define the contraction operation of tensors. Why this operation cannot be conducted on scalars and vectors?
   Answer: In tensor calculus, contraction of tensors means making two free indices of a given tensor or of two tensors involved in outer product identical by unifying their symbols. This will then be followed by performing summation over these repeated indices. For example, when we contract the rank-2 tensor A_i^j in nD space we obtain:
   $$A_i^i = A_1^1 + A_2^2 + \cdots + A_n^n$$
   Similarly, when we contract the j and k indices of the outer product $A_{ij}B^k$ in nD space we obtain:
   $$A_{ij}B^j = A_{i1}B^1 + A_{i2}B^2 + \cdots + A_{in}B^n$$
   From the above definition, we see that contraction requires two indices and hence it cannot be conducted on scalars and vectors since scalars have no index and vectors have only one index.

47. In reference to general coordinate systems, a single contraction operation is conducted on a tensor of type (m, n, w) where $m, n > 0$. What is the rank, the type and the weight of the contracted tensor?
   Answer: The rank is $m + n - 2$, the type is $(m - 1, n - 1, w)$ and the weight is w.

48. What is the condition that should be satisfied by the two tensor indices involved in a contraction operation assuming a general coordinate system? What about tensors in orthonormal Cartesian systems?
   Answer: In general coordinate systems the two tensor indices that are involved in a contraction operation should be of opposite variance type, i.e. one covariant and one contravariant. However, in orthonormal Cartesian systems the variance type is irrelevant, since there is no difference between covariant and contravariant types in these systems, and hence contraction can take place between any two indices in the same tensor term.

49. How many individual contraction operations can take place in a tensor of type (m, n, w) in a general coordinate system? Explain why.
   Answer: In general coordinate systems, $m \times n$ individual contraction operations can

take place in this tensor because we can have one contraction operation for each combination of one upper index and one lower index.

50. How many individual contraction operations can take place in a rank-n tensor in an orthonormal Cartesian coordinate system? Explain why.
    Answer: In orthonormal Cartesian systems, a rank-n tensor can have $\frac{n(n-1)}{2}$ possible individual contraction operations. The reason is that in these systems there is no difference between covariant and contravariant types and hence each one of the n indices can be contracted with each one of the remaining $(n-1)$ indices and therefore we should have $n(n-1)$ possible individual contraction operations. However, because the operation of contraction is independent of the order of the two contracted indices, since contracting i with j is the same as contracting j with i, then the number $n(n-1)$ should be reduced to half by dividing by 2 and hence we have $\frac{n(n-1)}{2}$ distinct contraction operations.

51. List all the possible single contraction operations that can take place in the tensor A^{ijk}_{lm}.
    Answer: Assuming a general coordinate system, we have: A^{ijk}_{im}, A^{ijk}_{li}, A^{ijk}_{jm}, A^{ijk}_{lj}, A^{ijk}_{km}, and A^{ijk}_{lk}.

52. List all the possible double contraction operations that can take place in the tensor A^{ij}_{kmn}.
    Answer: Assuming a general coordinate system, we have: A^{ij}_{ijn}, A^{ij}_{imj}, A^{ij}_{kij}, A^{ij}_{jin}, A^{ij}_{jmi} and A^{ij}_{kji}.

53. Give examples of contraction operation from matrix algebra.
    Answer: Taking the trace of a square matrix is an example of a contraction operation on a single tensor, while matrix product is an example of an operation that contains a contraction operation between two tensors.

54. Show that contracting a rank-n tensor results in a rank-$(n-2)$ tensor.
    Answer: The number of free indices in a rank-n tensor is n. On conducting a contraction operation on such a tensor 2 free indices will be consumed since they will become bound indices and hence the contracted tensor will become a rank-$(n-2)$ tensor.

55. Discuss inner multiplication of tensors as an operation composed of two more simple operations.
    Answer: Inner multiplication consists of an operation of outer multiplication on two non-scalar tensors followed by a contraction operation on two indices of the resultant product. Hence, inner multiplication can be seen as a composition of two more simple operations: outer multiplication and contraction.

56. Give common examples of inner product operation from linear algebra and vector calculus.
    Answer: An example of inner product operation from linear algebra is matrix multiplication. An example of inner product operation from vector calculus is the dot product of two vectors.

57. Why inner product operation is not commutative in general?
    Answer: Inner product operation is not commutative in general because the basis vectors of the basis tensor of the product are not commutative. For example, the inner

product operation of $\mathbf{A} = A_{ij}\mathbf{E}^i\mathbf{E}^j$ with $\mathbf{B} = B^{kl}\mathbf{E}_k\mathbf{E}_l$ which involves the indices j and k is:

$$\begin{aligned}
\mathbf{A} \cdot \mathbf{B} &= \left(A_{ij}\mathbf{E}^i\mathbf{E}^j\right) \cdot \left(B^{kl}\mathbf{E}_k\mathbf{E}_l\right) \\
&= A_{ij}B^{kl}\left(\mathbf{E}^i\mathbf{E}^j\right) \cdot \left(\mathbf{E}_k\mathbf{E}_l\right) \\
&= A_{ij}B^{kl}\mathbf{E}^i\left(\mathbf{E}^j \cdot \mathbf{E}_k\right)\mathbf{E}_l \\
&= A_{ij}B^{kl}\mathbf{E}^i\delta^j_k\mathbf{E}_l \\
&= A_{ij}B^{jl}\mathbf{E}^i\mathbf{E}_l
\end{aligned}$$

while the inner product operation of $\mathbf{B}$ with $\mathbf{A}$ which involves the same indices is:

$$\begin{aligned}
\mathbf{B} \cdot \mathbf{A} &= \left(B^{kl}\mathbf{E}_k\mathbf{E}_l\right) \cdot \left(A_{ij}\mathbf{E}^i\mathbf{E}^j\right) \\
&= B^{kl}A_{ij}\left(\mathbf{E}_k\mathbf{E}_l\right) \cdot \left(\mathbf{E}^i\mathbf{E}^j\right) \\
&= B^{kl}A_{ij}\mathbf{E}_l\left(\mathbf{E}_k \cdot \mathbf{E}^j\right)\mathbf{E}^i \\
&= B^{kl}A_{ij}\mathbf{E}_l\delta^j_k\mathbf{E}^i \\
&= B^{jl}A_{ij}\mathbf{E}_l\mathbf{E}^i
\end{aligned}$$

Now, since $\mathbf{E}^i\mathbf{E}_l \neq \mathbf{E}_l\mathbf{E}^i$ then the two operations are different and hence inner product operation is not commutative.

58. Complete the following equations where $\mathbf{A}$ and $\mathbf{B}$ are rank-2 tensors of opposite variance type:
$$\mathbf{A} : \mathbf{B} =? \qquad \mathbf{A} \cdot\cdot \mathbf{B} =?$$

Answer: If we use the same tensors of the previous question then we have:

$$\begin{aligned}
\mathbf{A} : \mathbf{B} &= \left(A_{ij}\mathbf{E}^i\mathbf{E}^j\right) : \left(B^{kl}\mathbf{E}_k\mathbf{E}_l\right) \\
&= A_{ij}B^{kl}\left(\mathbf{E}^i \cdot \mathbf{E}_k\right)\left(\mathbf{E}^j \cdot \mathbf{E}_l\right) \\
&= A_{ij}B^{kl}\delta^i_k\delta^j_l \\
&= A_{ij}B^{ij}
\end{aligned}$$

$$\begin{aligned}
\mathbf{A} \cdot\cdot \mathbf{B} &= \left(A_{ij}\mathbf{E}^i\mathbf{E}^j\right) \cdot\cdot \left(B^{kl}\mathbf{E}_k\mathbf{E}_l\right) \\
&= A_{ij}B^{kl}\left(\mathbf{E}^i \cdot \mathbf{E}_l\right)\left(\mathbf{E}^j \cdot \mathbf{E}_k\right) \\
&= A_{ij}B^{kl}\delta^i_l\delta^j_k \\
&= A_{ij}B^{ji}
\end{aligned}$$

59. Write $\mathbf{ab} : \mathbf{cd}$ in component form assuming a 3D Cartesian system. Repeat this with $\mathbf{ab} \cdot\cdot \mathbf{cd}$.

Answer: We have:

$$\begin{aligned}
\mathbf{ab} : \mathbf{cd} &= (\mathbf{a} \cdot \mathbf{c})(\mathbf{b} \cdot \mathbf{d}) \\
&= (a_1c_1 + a_2c_2 + a_3c_3)(b_1d_1 + b_2d_2 + b_3d_3)
\end{aligned}$$

$$
\begin{aligned}
&= a_1c_1b_1d_1 + a_2c_2b_1d_1 + a_3c_3b_1d_1 + \\
&\quad a_1c_1b_2d_2 + a_2c_2b_2d_2 + a_3c_3b_2d_2 + \\
&\quad a_1c_1b_3d_3 + a_2c_2b_3d_3 + a_3c_3b_3d_3
\end{aligned}
$$

$$
\begin{aligned}
\mathbf{ab}\cdot\cdot\mathbf{cd} &= (\mathbf{a}\cdot\mathbf{d})(\mathbf{b}\cdot\mathbf{c}) \\
&= (a_1d_1 + a_2d_2 + a_3d_3)(b_1c_1 + b_2c_2 + b_3c_3) \\
&= a_1d_1b_1c_1 + a_2d_2b_1c_1 + a_3d_3b_1c_1 + \\
&\quad a_1d_1b_2c_2 + a_2d_2b_2c_2 + a_3d_3b_2c_2 + \\
&\quad a_1d_1b_3c_3 + a_2d_2b_3c_3 + a_3d_3b_3c_3
\end{aligned}
$$

60. Why the operation of inner multiplication of tensors results in a tensor?
    Answer: Because inner product operation is synthesized from an outer product operation followed by a contraction operation and both these operations on tensors produce tensors. So, if we start from two tensors and subject them to outer multiplication we obtain a tensor, and when we subject this tensor to contraction we will also get a tensor (which is the final result of the inner multiplication). Accordingly, the operation of inner multiplication of tensors should produce a tensor.

61. We have: $\mathbf{A} = A^i\mathbf{E}_i$, $\mathbf{B} = B_j\mathbf{E}^j$ and $\mathbf{C} = C^m_n\mathbf{E}_m\mathbf{E}^n$. Find the following tensor products: $\mathbf{AB}$, $\mathbf{AC}$ and $\mathbf{BC}$.
    Answer:
$$
\begin{aligned}
\mathbf{AB} &= A^i B_j \mathbf{E}_i \mathbf{E}^j \\
\mathbf{AC} &= A^i C^m_n \mathbf{E}_i \mathbf{E}_m \mathbf{E}^n \\
\mathbf{BC} &= B_j C^m_n \mathbf{E}^j \mathbf{E}_m \mathbf{E}^n
\end{aligned}
$$

62. Referring to the tensors in question 61, find the following dot products: $\mathbf{B}\cdot\mathbf{B}$, $\mathbf{C}\cdot\mathbf{A}$ and $\mathbf{C}\cdot\mathbf{B}$.
    Answer:
$$
\begin{aligned}
\mathbf{B}\cdot\mathbf{B} &= (B_j\mathbf{E}^j)\cdot(B^k\mathbf{E}_k) \\
&= B_j B^k (\mathbf{E}^j \cdot \mathbf{E}_k) \\
&= B_j B^k \delta^j_k \\
&= B_j B^j
\end{aligned}
$$

$$
\begin{aligned}
\mathbf{C}\cdot\mathbf{A} &= (C^m_n\mathbf{E}_m\mathbf{E}^n)\cdot(A^i\mathbf{E}_i) \\
&= C^m_n A^i \mathbf{E}_m (\mathbf{E}^n \cdot \mathbf{E}_i) \\
&= C^m_n A^i \mathbf{E}_m \delta^n_i \\
&= C^m_i A^i \mathbf{E}_m
\end{aligned}
$$

$$
\mathbf{C}\cdot\mathbf{B} = (C^m_n\mathbf{E}_m\mathbf{E}^n)\cdot(B_j\mathbf{E}^j)
$$

3   TENSORS   51

$$
\begin{aligned}
&= C^m{}_n B_j \left(\mathbf{E}_m \cdot \mathbf{E}^j \right) \mathbf{E}^n \\
&= C^m{}_n B_j \delta^j_m \mathbf{E}^n \\
&= C^j{}_n B_j \mathbf{E}^n
\end{aligned}
$$

63. Define permutation of tensors giving an example of this operation from matrix algebra.
    Answer: Permutation of tensors is the operation of exchanging the position of two free indices. An example of this operation from matrix algebra is taking the transpose of a matrix by exchanging its rows and columns.
64. State the quotient rule of tensors in words and in a formal mathematical form.
    Answer: The essence of the quotient rule of tensors is that if the inner product of a suspected tensor by a known tensor is a tensor then the suspect is a tensor. Mathematically, if $\mathbf{A}$ is a suspected tensor and $\mathbf{B}$ and $\mathbf{C}$ are known tensors and we have:

    $$\mathbf{A} \cdot \mathbf{B} = \mathbf{C}$$

    then $\mathbf{A}$ is a tensor.
65. Why the quotient rule is usually used in tensor tests instead of applying the transformation rules?
    Answer: Because using the quotient rule is generally more convenient and requires less work than applying the transformation rules. Moreover, in some cases the quotient rule does not require actual work when we know, from our past experience, that a relation like the one seen in the previous question is already satisfied by our tensors so all we need from the quotient rule is to draw the conclusion that the suspected tensor is a tensor indeed.
66. Outline the similarities and differences between the three main forms of tensor representation, i.e. covariant, contravariant and physical.
    Answer: Some valid points are:
    • All these forms are legitimate representations of tensors and hence any tensor can be represented covariantly, contravariantly or physically without affecting the properties of the tensor, i.e. all these different forms represent the same tensor.
    • The components of the covariant and contravariant forms may have different physical dimensions (or even dimensionless), but the components of the physical form have the same physical dimension. This also applies to the basis vectors of these representations.
    • The basis vectors in the physical representation are normalized and dimensionless, and this may not be the case in the covariant and contravariant representations.
67. Define, mathematically, the physical basis vectors, $\hat{\mathbf{E}}_i$ and $\hat{\mathbf{E}}^i$, in terms of the covariant and contravariant basis vectors, $\mathbf{E}_i$ and $\mathbf{E}^i$.
    Answer: We have:

    $$\hat{\mathbf{E}}_i = \frac{\mathbf{E}_i}{|\mathbf{E}_i|} \qquad \text{and} \qquad \hat{\mathbf{E}}^i = \frac{\mathbf{E}^i}{|\mathbf{E}^i|}$$

    where $|\mathbf{E}_i|$ and $|\mathbf{E}^i|$ are the magnitudes of $\mathbf{E}_i$ and $\mathbf{E}^i$ respectively.

68. Correct, if necessary, the following relation: $\hat{A}^{ikn}_{jm} = \frac{h_i h_j h_n}{h_k h_m} A^{ikn}_{jm}$ (no sum on any index) where **A** is a tensor in an orthogonal coordinate system.
    Answer: The correct form of this relation is:
    $$\hat{A}^{ikn}_{jm} = \frac{h_i h_k h_n}{h_j h_m} A^{ikn}_{jm} \qquad \text{(no sum on any index)}$$

69. Why the normalized covariant, contravariant and physical basis vectors are identical in orthogonal coordinate systems?
    Answer: Because in orthogonal coordinate systems the corresponding covariant and contravariant basis vectors have the same direction and hence when they are normalized they become identical. So, the normalized basis vector sets of all these representations are the same.

70. What is the physical significance of being able to transform one type of tensors to other types as well as transforming between different coordinate systems?
    Answer: The physical significance is that the tensor as a mathematical and physical entity does not change by transforming from one type to another (e.g. covariant to contravariant) and hence it possesses the same properties (i.e. it is invariant in this sense) regardless of the type by which it is represented. Moreover, the tensor is invariant across all coordinates systems and hence its real properties will not change by transforming from one system to another. This invariant nature of tensors makes them very valuable tool in formulating the laws of physics which should be invariant across all representations and across all coordinate systems.

71. Why the physical representation of tensors is usually preferred in the scientific applications of tensor calculus?
    Answer: Because the physical representation of tensors is a standardized form where all the components have the same physical dimension while all the basis vectors are normalized and dimensionless. This uniformity facilitates the management and comprehension of tensors in the scientific applications of tensor calculus.

72. Give a few common examples of physical representation of tensors in mathematical and scientific applications.
    Answer: Many examples can be found in fluid or continuum mechanics or general relativity for instance where Cartesian or cylindrical or spherical coordinate systems with normalized dimensionless basis vectors are used to represent and formulate tensors (e.g. stress tensor) in physical form (refer to § 7 in the book).

73. What is the advantage of representing the physical components of a tensor (e.g. **A**) by the symbol of the tensor with subscripts denoting the coordinates of the employed coordinate system, e.g. (A_r, A_θ, A_ϕ) in spherical coordinate systems?
    Answer: One advantage is that the employed coordinate system to which the tensor is referred can be easily inferred from the notation. Another advantage is that the coordinates to which the components are referred will not be confused if the order of coordinates is unclear or it is susceptible to change, unlike using numbers for example (e.g. A_1, A_2, A_3) to label the coordinates.

Chapter 4
Special Tensors

1. What is special about the Kronecker delta, the permutation and the metric tensors and why they deserve special attention?
   Answer: Some of the special characteristics of these tensors that qualify them for special attention are:
   - They are part of the theory of tensor calculus itself and hence they are present almost everywhere in tensor calculus and its applications.
   - They enter in essential definitions and operations of tensor calculus.
   - They have very distinct mathematical properties that distinguish them from other tensors (refer to the book for details).

2. Give detailed definition, in words and in symbols, of the ordinary Kronecker delta tensor in an nD space.
   Answer: The ordinary Kronecker delta tensor, also known as the unit tensor, is a rank-2 numeric, absolute, symmetric, constant, isotropic tensor in all dimensions. It is defined in its covariant form as:

   $$\delta_{ij} = \begin{cases} 1 & (i = j) \\ 0 & (i \neq j) \end{cases} \qquad (i, j = 1, 2, \ldots n)$$

   where n is the space dimension, and hence it can be considered as the identity tensor or matrix. The above definition similarly applies to the contravariant and mixed forms of this tensor (i.e. δ^{ij} and δ^i_j).

3. List and discuss all the main characteristics (e.g. symmetry) of the ordinary Kronecker delta tensor.
   Answer: The main characteristics of this tensor are:
   - It is a rank-2 tensor and hence it possesses n^2 components in an nD space.
   - It is numeric tensor and hence the values of its components are 1 and 0 in any coordinate system.
   - The value of any particular component (e.g. δ_{12}) of this tensor is the same in any coordinate system and hence it is constant tensor in this sense.
   - Its components have identical values in all variance types, i.e. $\delta_{ij} = \delta^{ij} = \delta^i_j$.
   - It is symmetric tensor for both variance types and hence $\delta_{ij} = \delta_{ji}$ and $\delta^{ij} = \delta^{ji}$.
   - It is absolute tensor and hence its weight w is zero.
   - It possesses the above properties in any space dimension and in any coordinate system.

4. Write the matrix that represents the ordinary Kronecker delta tensor in a 4D space.

$$\begin{bmatrix} 1 & 0 & 0 & 0 \\ 0 & 1 & 0 & 0 \\ 0 & 0 & 1 & 0 \\ 0 & 0 & 0 & 1 \end{bmatrix}$$

5. Do we violate the rules of tensor indices when we write: $\delta_{ij} = \delta^{ij} = \delta^i_j = \delta^{\,j}_i$?
   Answer: No, because these equalities belong to the values of the components of the Kronecker delta tensor and hence they are essentially scalar equalities and not tensor equalities. We may claim that the indices in these equalities are labels to the individual components rather than tensor indices that can vary over their range.

6. Explain the following statement: "The ordinary Kronecker delta tensor is conserved under all proper and improper coordinate transformations". What is the relation between this and the property of isotropy of this tensor?
   Answer: This statement means that the components of the ordinary Kronecker delta tensor are constant and hence they do not change under proper or improper coordinate transformations. Being conserved in this sense is stronger than being isotropic because the former applies to both proper and improper transformations while the latter applies by definition only to proper transformations.

7. List and discuss all the main characteristics (e.g. anti-symmetry) of the permutation tensor.
   Answer: The main characteristics of the permutation tensor are:
   • It is numeric tensor and hence the values of its components are -1, 1 and 0 in all coordinate systems.
   • The value of any particular component (e.g. ϵ_{312}) of this tensor is the same in any coordinate system and hence it is constant tensor in this sense.
   • It is relative tensor of weight -1 for its covariant form and $+1$ for its contravariant form.
   • It is isotropic tensor since its components are conserved under proper transformations.
   • It is totally anti-symmetric in each pair of its indices for both variance types, i.e. it changes sign on swapping any two of its indices.
   • It is pseudo tensor since it acquires a minus sign under improper orthogonal transformation of coordinates.
   • The permutation tensor of any rank has only one independent non-vanishing component because all the non-zero components of this tensor are either $+1$ or -1.
   • The rank-n permutation tensor possesses $n!$ non-zero components which is the number of the non-repetitive permutations of its indices.
   • The rank and the dimension of the permutation tensor are identical and hence in an nD space it is of rank-n and therefore it has n^n components.

8. What are the other names used to label the permutation tensor?
   Answer: The permutation tensor is also known as the Levi-Civita tensor, the anti-symmetric tensor and the alternating tensor.

9. Why the rank and the dimension of the permutation tensor are the same? Accordingly, what is the number of components of the rank-2, rank-3 and rank-4 permutation ten-

4 SPECIAL TENSORS   55

sors?
Answer: There are two main defining properties of the permutation tensor:
(a) It is very abstract mathematical entity that has no association with any physical property of the space and hence if it is associated with any property of the space it should be the dimensionality of the space.
(b) It is about permuting or alternating something and hence it should be about permuting or alternating something related to the dimensionality of the space.
Accordingly, if each independent dimension of the space is identified by an independent free index then the permutation tensor (whose function is to permute these indices that refer to the dimensions of the space) of an nD space should have n free indices and hence it is of rank-n, i.e. the rank and the dimension of the permutation tensor are the same, as claimed.
The number of components of a rank-r tensor in an nD space is n^r and hence the number of components of the permutation tensor in an nD space is n^n. Therefore, the number of components of the rank-2, rank-3 and rank-4 permutation tensors are respectively: $2^2 = 4$, $3^3 = 27$ and $4^4 = 256$.

10. Why the permutation tensor of any rank has only one independent non-vanishing component?
    Answer: By definition, the components of the permutation tensor of any rank are either 0 or +1 or −1. So, if we exclude the vanishing components (i.e. 0) then all the remaining non-vanishing components are of unity magnitude (i.e. either +1 or −1) and hence we have only one independent non-vanishing component (i.e. all these components can be obtained from a single non-vanishing component by at most a reversal of sign and hence they are not independent).

11. Prove that the rank-n permutation tensor possesses $n!$ non-zero components.
    Answer: By definition, all the components of the permutation tensor with repetitive indices are zero while all the components of the permutation tensor with non-repetitive indices are non-zero (i.e. either +1 or −1). Therefore, the non-zero components of the permutation tensor of any rank are restricted to the non-repetitive permutations of the indices. Now, for rank-n permutation tensor we have n distinct indices; moreover the number of non-repetitive permutations of n objects is $n!$ (i.e. n possible selections of n distinct objects times $n-1$ possible selections of the remaining $n-1$ objects $\cdots$ times 2 times 1). Hence, the number of non-zero components of the rank-n permutation tensor is $n!$.

12. Why the permutation tensor is totally anti-symmetric?
    Answer: By definition, the exchange of any two indices of the permutation tensor leads to reversal of sign and hence it is totally anti-symmetric according to the definition of totally anti-symmetric tensor.

13. Give the inductive mathematical definition of the components of the permutation tensor of rank-n.

Answer: The components of the rank-n permutation tensor are defined inductively as:

$$\epsilon_{i_1 i_2 \ldots i_n} = \epsilon^{i_1 i_2 \ldots i_n} = \begin{cases} 1 & (i_1, i_2, \ldots, i_n \text{ is even permutation of } 1, 2, \ldots, n) \\ -1 & (i_1, i_2, \ldots, i_n \text{ is odd permutation of } 1, 2, \ldots, n) \\ 0 & (\text{repeated index}) \end{cases}$$

14. State the most simple analytical mathematical definition of the components of the permutation tensor of rank-n.
    Answer: It is:
    $$\epsilon_{a_1 a_2 \cdots a_n} = \epsilon^{a_1 a_2 \cdots a_n} = \prod_{1 \le i < j \le n} \mathrm{sgn}\,(a_j - a_i)$$
    where sgn is the sign function.

15. Make a sketch of the array representing the rank-3 permutation tensor where the nodes of the array are marked with the symbols and values of the components of this tensor.
    Answer: The sketch should look like Figure 6.

16. Define, mathematically, the rank-n covariant and contravariant absolute permutation tensors, $\underline{\epsilon}_{i_1 \ldots i_n}$ and $\underline{\epsilon}^{i_1 \ldots i_n}$.
    Answer: They are:
    $$\begin{aligned} \underline{\epsilon}_{i_1 \ldots i_n} &= \sqrt{g}\, \epsilon_{i_1 \ldots i_n} & \text{(Covariant)} \\ \underline{\epsilon}^{i_1 \ldots i_n} &= \frac{1}{\sqrt{g}} \epsilon^{i_1 \ldots i_n} & \text{(Contravariant)} \end{aligned}$$

    where the indexed ϵ and $\underline{\epsilon}$ are respectively the relative and absolute permutation tensors of the given type, and g is the determinant of the covariant metric tensor g_{ij}.

17. Show that ϵ_{ijk} is a relative tensor of weight -1 and ϵ^{ijk} is a relative tensor of weight $+1$.
    Answer: We start from the definition of the determinant of a rank-2 tensor in 3D space which we gave in this chapter of the book, that is:
    $$\det(\mathbf{A}) = \frac{1}{3!} \epsilon^{ijk} \epsilon_{lmn} A_i^l A_j^m A_k^n$$

    Now, if we replace $\mathbf{A}$ with $\mathbf{J}$, where $\mathbf{J}$ stands for the Jacobian matrix, then the last equation can be written as:
    $$J = \frac{1}{3!} \epsilon^{ijk} \epsilon_{lmn} J_i^l J_j^m J_k^n$$
    where J is the Jacobian while $J_i^l = \frac{\partial u^l}{\partial \bar{u}^i}$ and J_j^m and J_k^n are defined similarly. Now, if we multiply both sides of the last equation with ϵ_{ijk} and note the identity: $\epsilon_{ijk}\epsilon^{ijk} = 3!$ which we gave in this chapter of the book, then the last equation will become:
    $$\begin{aligned} \epsilon_{ijk} J &= \epsilon_{lmn} J_i^l J_j^m J_k^n \\ \epsilon_{ijk} &= J^{-1} \epsilon_{lmn} J_i^l J_j^m J_k^n \end{aligned}$$

4 SPECIAL TENSORS

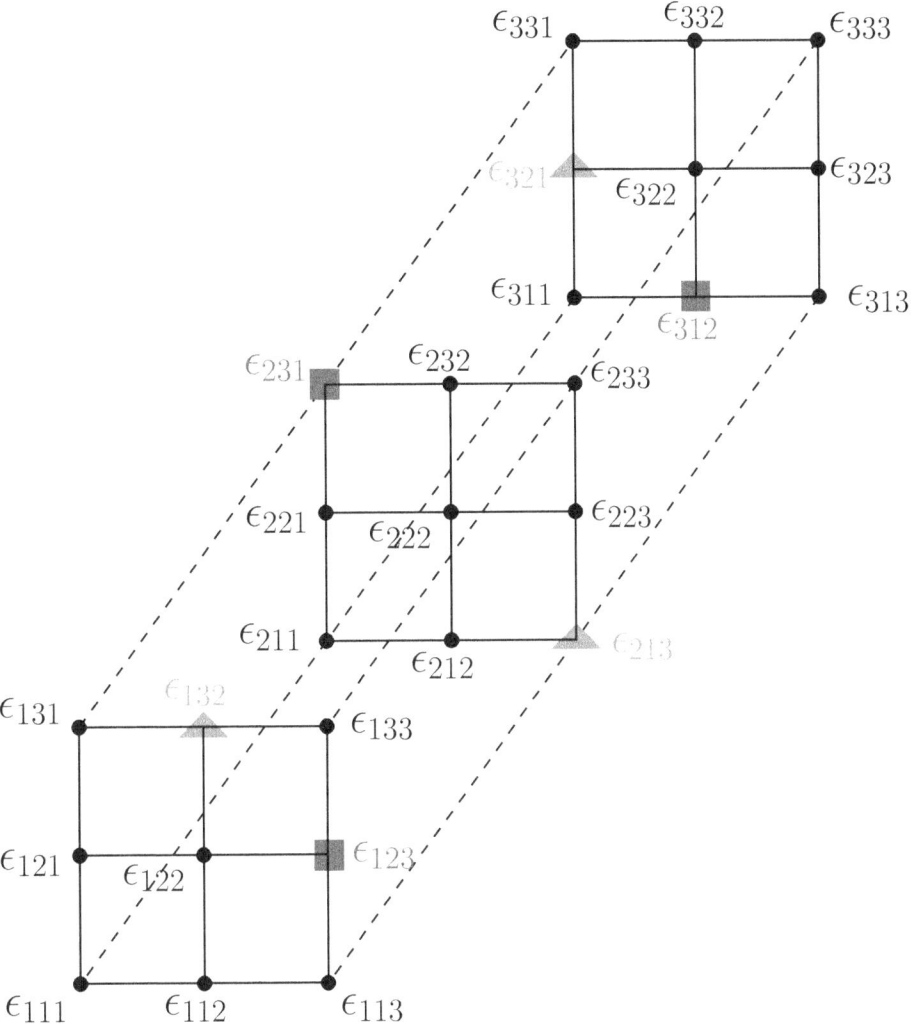

Figure 6: Graphical illustration of the rank-3 permutation tensor ϵ_{ijk} where the black circles represent the 0 components, the blue squares represent the 1 components and the red triangles represent the -1 components.

$$\epsilon_{ijk} = J^{-1}\epsilon_{lmn}\frac{\partial u^l}{\partial \bar{u}^i}\frac{\partial u^m}{\partial \bar{u}^j}\frac{\partial u^n}{\partial \bar{u}^k}$$

$$\bar{\epsilon}_{ijk} = J^{-1}\epsilon_{lmn}\frac{\partial u^l}{\partial \bar{u}^i}\frac{\partial u^m}{\partial \bar{u}^j}\frac{\partial u^n}{\partial \bar{u}^k}$$

where the last step is justified by the fact that the permutation tensor is conserved across all coordinate systems and hence $\epsilon_{ijk} = \bar{\epsilon}_{ijk}$.[14] As we see, the last equation is the transformation equation of a covariant rank-3 relative tensor of weight $w = -1$ from unbarred to barred systems and hence ϵ_{ijk} is a relative tensor of weight -1, as required.

---

[14] The reader should note that the over-bar refers to the barred system and hence it should not be confused with the under-bar which we use to label absolute permutation tensors.

If we repeat the above argument on the contravariant permutation tensor, then from the definition of the determinant we have:

$$\det(\mathbf{A}) = \frac{1}{3!} \epsilon^{ijk} \epsilon_{lmn} A_i^l A_j^m A_k^n$$

Now, if we replace $\mathbf{A}$ with $\mathbf{J}^{-1}$, where $\mathbf{J}^{-1}$ is the inverse Jacobian matrix, then the last equation will become:

$$J^{-1} = \frac{1}{3!} \epsilon^{ijk} \epsilon_{lmn} \left(J^{-1}\right)_i^l \left(J^{-1}\right)_j^m \left(J^{-1}\right)_k^n$$

where J^{-1} is the inverse Jacobian while $\left(J^{-1}\right)_i^l = \frac{\partial \bar{u}^l}{\partial u^i}$ and $\left(J^{-1}\right)_j^m$ and $\left(J^{-1}\right)_k^n$ are defined similarly. Now, if we multiply both sides of the last equation with ϵ^{lmn} and note the identity: $\epsilon_{lmn} \epsilon^{lmn} = 3!$ then the last equation will become:

$$\begin{aligned}
\epsilon^{lmn} J^{-1} &= \epsilon^{ijk} \left(J^{-1}\right)_i^l \left(J^{-1}\right)_j^m \left(J^{-1}\right)_k^n \\
\epsilon^{lmn} &= J^{+1} \epsilon^{ijk} \left(J^{-1}\right)_i^l \left(J^{-1}\right)_j^m \left(J^{-1}\right)_k^n \\
\epsilon^{lmn} &= J^{+1} \epsilon^{ijk} \frac{\partial \bar{u}^l}{\partial u^i} \frac{\partial \bar{u}^m}{\partial u^j} \frac{\partial \bar{u}^n}{\partial u^k} \\
\bar{\epsilon}^{lmn} &= J^{+1} \epsilon^{ijk} \frac{\partial \bar{u}^l}{\partial u^i} \frac{\partial \bar{u}^m}{\partial u^j} \frac{\partial \bar{u}^n}{\partial u^k}
\end{aligned}$$

where the last step is justified by the fact that the permutation tensor is conserved across all coordinate systems and hence $\epsilon^{lmn} = \bar{\epsilon}^{lmn}$. As we see, the last equation is the transformation equation of a contravariant rank-3 relative tensor of weight $w = +1$ from unbarred to barred systems and hence ϵ^{ijk} is a relative tensor of weight $+1$, as required.

18. Show that $\underline{\epsilon}_{i_1 \ldots i_n} = \sqrt{g}\, \epsilon_{i_1 \ldots i_n}$ and $\underline{\epsilon}^{i_1 \ldots i_n} = \frac{1}{\sqrt{g}} \epsilon^{i_1 \ldots i_n}$ are absolute tensors.

    Answer: If we follow the arguments of the last question, then we should get the following transformation equations:

$$\begin{aligned}
\bar{\epsilon}_{i_1 \ldots i_n} &= J^{-1} \epsilon_{j_1 \ldots j_n} \frac{\partial u^{j_1}}{\partial \bar{u}^{i_1}} \cdots \frac{\partial u^{j_n}}{\partial \bar{u}^{i_n}} \\
\bar{\epsilon}^{i_1 \ldots i_n} &= J^{+1} \epsilon^{j_1 \ldots j_n} \frac{\partial \bar{u}^{i_1}}{\partial u^{j_1}} \cdots \frac{\partial \bar{u}^{i_n}}{\partial u^{j_n}}
\end{aligned}$$

Now, it was shown earlier (see Exercise 18 of § 3) that the determinant of a rank-2 absolute covariant tensor is a relative scalar of weight 2. On applying this rule on the metric tensor g_{ij}, which is a rank-2 absolute covariant tensor, we get:

$$\begin{aligned}
|\bar{g}_{ij}| &= J^2 |g_{pq}| \\
\bar{g} &= J^2 g \\
\sqrt{\bar{g}} &= J \sqrt{g}
\end{aligned}$$

On multiplying the two sides of the first transformation equation by the two sides of the last equation, and dividing the two sides of the second transformation equation by the two sides of the last equation, we get:

$$\sqrt{\bar{g}}\,\bar{\epsilon}_{i_1\ldots i_n} = \sqrt{g}\,\epsilon_{j_1\ldots j_n}\frac{\partial u^{j_1}}{\partial \bar{u}^{i_1}}\cdots\frac{\partial u^{j_n}}{\partial \bar{u}^{i_n}}$$

$$\frac{\bar{\epsilon}^{i_1\ldots i_n}}{\sqrt{\bar{g}}} = \frac{\epsilon^{j_1\ldots j_n}}{\sqrt{g}}\frac{\partial \bar{u}^{i_1}}{\partial u^{j_1}}\cdots\frac{\partial \bar{u}^{i_n}}{\partial u^{j_n}}$$

The last two equations mean that

$$\sqrt{g}\,\epsilon_{j_1\ldots j_n} \equiv \underline{\epsilon}_{j_1\ldots j_n} \qquad \text{and} \qquad \frac{\epsilon^{j_1\ldots j_n}}{\sqrt{g}} \equiv \underline{\epsilon}^{j_1\ldots j_n}$$

transform as absolute tensors and hence they are absolute tensors, as required (noting the trivial difference in the labeling of indices with the labeling in the question).

19. Write $\epsilon^{i_1 i_2 \cdots i_n}\,\epsilon_{j_1 j_2 \cdots j_n}$ in its determinantal form in terms of the ordinary Kronecker delta.
    Answer:

    $$\epsilon^{i_1 i_2 \cdots i_n}\,\epsilon_{j_1 j_2 \cdots j_n} = \begin{vmatrix} \delta^{i_1}_{j_1} & \delta^{i_1}_{j_2} & \cdots & \delta^{i_1}_{j_n} \\ \delta^{i_2}_{j_1} & \delta^{i_2}_{j_2} & \cdots & \delta^{i_2}_{j_n} \\ \vdots & \vdots & \ddots & \vdots \\ \delta^{i_n}_{j_1} & \delta^{i_n}_{j_2} & \cdots & \delta^{i_n}_{j_n} \end{vmatrix}$$

20. Prove the following identity: $\epsilon^{i_1 i_2 \cdots i_n}\,\epsilon_{i_1 i_2 \cdots i_n} = n!$.
    Answer: According to the summation convention, $\epsilon^{i_1 i_2 \cdots i_n}\,\epsilon_{i_1 i_2 \cdots i_n}$ is the sum of products of $\epsilon^{i_1 i_2 \cdots i_n}$ and $\epsilon_{i_1 i_2 \cdots i_n}$ where both these should be 0 or $+1$ or -1 and hence the terms of this sum is either 0 or $+1$. Now, the number of terms of this sum is equal to the number of all permutations of $i_1, i_2, \cdots i_n$. However, all the permutations with repetitive indices are zero and hence only the terms of non-repetitive indices are equal to $+1$. Because there are $n!$ non-repetitive permutations of $i_1, i_2, \cdots i_n$ then we have $n!$ terms of $+1$ and hence their sum is equal to $n!$, as required.

21. State a mathematical relation representing the use of the ordinary Kronecker delta tensor as an index replacement operator.
    Answer: For example, in the relation $\delta^j_i A^{kl}_j = A^{kl}_i$ the ordinary Kronecker delta tensor replaced the covariant index j of the tensor A^{kl}_j with the index i and the result is changing A^{kl}_j to A^{kl}_i. Similarly, in the relation $\delta^i_j B^j_k = B^i_k$ the ordinary Kronecker delta tensor replaced the contravariant index j of the tensor B^j_k with the index i and the result is changing B^j_k to B^i_k.

22. Prove the following relation inductively by starting from writing it in an expanded form in a 2D space: $\delta^i_i = n$.
    Answer: In a 2D space we have:

    $$\delta^i_i = \delta^1_1 + \delta^2_2 = 1 + 1 = 2 = n$$

    Moreover, if in an nD space this relation is true, that is:

    $$\delta^i_i = \delta^1_1 + \delta^2_2 + \cdots + \delta^n_n = 1 + 1 + \cdots + 1 = n$$

then in an $(n+1)$D space it should also be true because we then have:

$$\delta_i^i = \delta_1^1 + \delta_2^2 + \cdots + \delta_n^n + \delta_{n+1}^{n+1} = 1 + 1 + \cdots + 1 + 1 = n + 1$$

Hence, by mathematical induction the relation $\delta_i^i = n$ is true in any space of any dimension.

23. Repeat exercise 22 with the relation: $u^i{}_{,j} = \delta_j^i$ using a matrix form.
    Answer: In a 2D space we have:

$$[u^i{}_{,j}] = \begin{bmatrix} \frac{\partial u^1}{\partial u^1} & \frac{\partial u^1}{\partial u^2} \\ \frac{\partial u^2}{\partial u^1} & \frac{\partial u^2}{\partial u^2} \end{bmatrix} = \begin{bmatrix} 1 & 0 \\ 0 & 1 \end{bmatrix} = [\delta_j^i]$$

Moreover, if in an nD space this relation is true, that is:

$$[u^i{}_{,j}] = \begin{bmatrix} \frac{\partial u^1}{\partial u^1} & \frac{\partial u^1}{\partial u^2} & \cdots & \frac{\partial u^1}{\partial u^n} \\ \frac{\partial u^2}{\partial u^1} & \frac{\partial u^2}{\partial u^2} & \cdots & \frac{\partial u^2}{\partial u^n} \\ \vdots & \vdots & \ddots & \vdots \\ \frac{\partial u^n}{\partial u^1} & \frac{\partial u^n}{\partial u^2} & \cdots & \frac{\partial u^n}{\partial u^n} \end{bmatrix} = \begin{bmatrix} 1 & 0 & \cdots & 0 \\ 0 & 1 & \cdots & 0 \\ \vdots & \vdots & \ddots & \vdots \\ 0 & 0 & \cdots & 1 \end{bmatrix} = [\delta_j^i]$$

then in an $(n+1)$D space it should also be true because we then have:

$$[u^i{}_{,j}] = \begin{bmatrix} \frac{\partial u^1}{\partial u^1} & \frac{\partial u^1}{\partial u^2} & \cdots & \frac{\partial u^1}{\partial u^n} & \frac{\partial u^1}{\partial u^{n+1}} \\ \frac{\partial u^2}{\partial u^1} & \frac{\partial u^2}{\partial u^2} & \cdots & \frac{\partial u^2}{\partial u^n} & \frac{\partial u^2}{\partial u^{n+1}} \\ \vdots & \vdots & \ddots & \vdots & \vdots \\ \frac{\partial u^n}{\partial u^1} & \frac{\partial u^n}{\partial u^2} & \cdots & \frac{\partial u^n}{\partial u^n} & \frac{\partial u^n}{\partial u^{n+1}} \\ \frac{\partial u^{n+1}}{\partial u^1} & \frac{\partial u^{n+1}}{\partial u^2} & \cdots & \frac{\partial u^{n+1}}{\partial u^n} & \frac{\partial u^{n+1}}{\partial u^{n+1}} \end{bmatrix} = \begin{bmatrix} 1 & 0 & \cdots & 0 & 0 \\ 0 & 1 & \cdots & 0 & 0 \\ \vdots & \vdots & \ddots & \vdots & \vdots \\ 0 & 0 & \cdots & 1 & 0 \\ 0 & 0 & \cdots & 0 & 1 \end{bmatrix} = [\delta_j^i]$$

Hence, by mathematical induction the relation $u^i{}_{,j} = \delta_j^i$ is true in any space of any dimension.

24. Justify the following relation assuming an orthonormal Cartesian system: $\partial_i x_j = \partial_j x_i$.
    Answer: Due to the fact that the coordinates are independent of each other, plus the fact that the covariant and contravariant types are the same in orthonormal Cartesian coordinate systems, we have:

$$\partial_i x_j \equiv \frac{\partial x_j}{\partial x_i} = \delta_{ji} = \delta_{ij} = \frac{\partial x_i}{\partial x_j} \equiv \partial_j x_i$$

where the middle equality (i.e. $\delta_{ji} = \delta_{ij}$) is justified by the symmetry of the Kronecker delta tensor.

25. Justify the following relations where the indexed **e** are orthonormal vectors:

$$\mathbf{e}_i \cdot \mathbf{e}_j = \delta_{ij} \qquad\qquad \mathbf{e}_i \mathbf{e}_j : \mathbf{e}_k \mathbf{e}_l = \delta_{ik} \delta_{jl}$$

Answer: Regarding the relation $\mathbf{e}_i \cdot \mathbf{e}_j = \delta_{ij}$, because $\mathbf{e}_i$ and $\mathbf{e}_j$ are orthogonal when $i \neq j$ then $\mathbf{e}_i \cdot \mathbf{e}_j = 0$ when $i \neq j$, and because they are normalized then $\mathbf{e}_i \cdot \mathbf{e}_j = 1$

4 SPECIAL TENSORS   61

when $i = j$. These two conditions are equivalent to the conditions of the Kronecker delta tensor (i.e. $\delta_{ij} = 0$ when $i \neq j$ and $\delta_{ij} = 1$ when $i = j$) and hence $\mathbf{e}_i \cdot \mathbf{e}_j = \delta_{ij}$, as required.

Regarding the relation $\mathbf{e}_i \mathbf{e}_j : \mathbf{e}_k \mathbf{e}_l = \delta_{ik}\delta_{jl}$, we have:

$$\mathbf{e}_i \mathbf{e}_j : \mathbf{e}_k \mathbf{e}_l = (\mathbf{e}_i \cdot \mathbf{e}_k)(\mathbf{e}_j \cdot \mathbf{e}_l) = \delta_{ik}\delta_{jl}$$

where the first equality is justified by the definition of the double inner product (i.e. :) which is given in the book, while the second equality is justified by the answer of the first part of this question.

26. Show that $\delta_i^j \delta_j^k \delta_k^i = n$.

    Answer: We have:

    $$\delta_i^j \delta_j^k \delta_k^i = \delta_i^k \delta_k^i = \delta_i^i = n$$

    where in the first and second equalities the Kronecker delta is acting as an index replacement operator while the third equality is based on the identity $\delta_i^i = n$ which was proved earlier (see Exercise 22).

27. Write the determinantal array form of $\epsilon^{ij}\epsilon_{kl}$ outlining the pattern of the tensor indices in their relation to the indices of the rows and columns of the determinant array.

    Answer:

    $$\epsilon^{ij}\epsilon_{kl} = \begin{vmatrix} \delta_k^i & \delta_l^i \\ \delta_k^j & \delta_l^j \end{vmatrix} = \delta_k^i \delta_l^j - \delta_l^i \delta_k^j$$

    The pattern of the indices in the determinant array of this identity is that the indices of the first ϵ provide the indices for the rows as the upper indices of the deltas while the indices of the second ϵ provide the indices for the columns as the lower indices of the deltas.

28. Prove the following identity using a truth table: $\epsilon^{ij}\epsilon_{kl} = \delta_k^i \delta_l^j - \delta_l^i \delta_k^j$.

    Answer: The truth table is given in the text where on comparing the column of $\epsilon^{ij}\epsilon_{kl}$ with the column of $\delta_k^i \delta_l^j - \delta_l^i \delta_k^j$ we see that the corresponding entries have identical values and hence $\epsilon^{ij}\epsilon_{kl} = \delta_k^i \delta_l^j - \delta_l^i \delta_k^j$.

29. Prove the following identities justifying each step in your proofs:

    $$\epsilon^{il}\epsilon_{kl} = \delta_k^i \qquad\qquad \epsilon^{ijk}\epsilon_{lmk} = \delta_l^i \delta_m^j - \delta_m^i \delta_l^j$$

    Answer: Regarding the identity $\epsilon^{il}\epsilon_{kl} = \delta_k^i$, we have:

    $$\begin{aligned}\epsilon^{il}\epsilon_{kl} &= \delta_k^i \delta_l^l - \delta_l^i \delta_k^l \\ &= 2\delta_k^i - \delta_l^i \delta_k^l \\ &= 2\delta_k^i - \delta_k^i \\ &= \delta_k^i\end{aligned}$$

    where the first line is based on the identity of the previous question with the replacement of j with l, the second line is justified by the previously proved identity $\delta_i^i = n$ (see Exercise 22) where $n = 2$ since the dimension and rank of ϵ are identical, and the third

line is based on the action of the Kronecker delta as an index replacement operator. Regarding the identity $\epsilon^{ijk}\epsilon_{lmk} = \delta^i_l \delta^j_m - \delta^i_m \delta^j_l$, we have:

$$
\begin{aligned}
\epsilon^{ijk}\epsilon_{lmk} &= \delta^i_l \delta^j_m \delta^k_k + \delta^i_m \delta^j_k \delta^k_l + \delta^i_k \delta^j_l \delta^k_m - \delta^i_l \delta^j_k \delta^k_m - \delta^i_m \delta^j_l \delta^k_k - \delta^i_k \delta^j_m \delta^k_l \\
&= 3\delta^i_l \delta^j_m + \delta^i_m \delta^j_k \delta^k_l + \delta^i_k \delta^j_l \delta^k_m - \delta^i_l \delta^j_k \delta^k_m - 3\delta^i_m \delta^j_l - \delta^i_k \delta^j_m \delta^k_l \\
&= 3\delta^i_l \delta^j_m + \delta^i_m \delta^j_l + \delta^i_m \delta^j_l - \delta^i_l \delta^j_m - 3\delta^i_m \delta^j_l - \delta^i_l \delta^j_m \\
&= \delta^i_l \delta^j_m - \delta^i_m \delta^j_l
\end{aligned}
$$

where the first line is justified by the identity:

$$
\epsilon^{ijk}\epsilon_{lmn} = \begin{vmatrix} \delta^i_l & \delta^i_m & \delta^i_n \\ \delta^j_l & \delta^j_m & \delta^j_n \\ \delta^k_l & \delta^k_m & \delta^k_n \end{vmatrix}
$$

with the replacement of n with k, the second line is justified by the previously proved identity $\delta^i_i = n$ (see Exercise 22) where $n = 3$ since the dimension and rank of ϵ are identical, and the third line is based on the action of the Kronecker delta as an index replacement operator.

30. Prove the following identities using other more general identities:

$$
\epsilon^{ijk}\epsilon_{ljk} = 2\delta^i_l \qquad \epsilon^{ijk}\epsilon_{ijk} = 6
$$

Answer: Regarding the first identity, we have:

$$
\begin{aligned}
\epsilon^{ijk}\epsilon_{ljk} &= \delta^i_l \delta^j_j - \delta^i_j \delta^j_l \\
&= 3\delta^i_l - \delta^i_j \delta^j_l \\
&= 3\delta^i_l - \delta^i_l \\
&= 2\delta^i_l
\end{aligned}
$$

where in the first line we are using the identity $\epsilon^{ijk}\epsilon_{lmk} = \delta^i_l \delta^j_m - \delta^i_m \delta^j_l$ of the previous question with the contraction of j and m, the second line is justified by the previously proved identity $\delta^i_i = n$ (see Exercise 22) where $n = 3$ since the dimension and rank of ϵ are identical, and the third line is based on the action of the Kronecker delta as an index replacement operator.

Regarding the second identity, we have:

$$
\epsilon^{ijk}\epsilon_{ijk} = 2\delta^i_i = 2 \times 3 = 6
$$

where in the first equality we are using the first identity of this question (i.e. $\epsilon^{ijk}\epsilon_{ljk} = 2\delta^i_l$) with the contraction of i and l, while in the second equality we are using the previously proved identity $\delta^i_i = n$ with $n = 3$.

31. Outline the similarities and differences between the ordinary Kronecker delta tensor and the generalized Kronecker delta tensor.
    Answer: There are many valid points that can be given in the answer of this question;

4 SPECIAL TENSORS

some of these are:
- Both Kronecker deltas are constant, numeric, absolute, isotropic tensors in all dimensions.
- Both Kronecker deltas have close relation with the permutation tensor. They can both be defined in terms of the permutation tensor and they share many identities with this tensor.
- While ordinary Kronecker delta is a rank-2 tensor in any space, the generalized Kronecker delta is a rank-$2n$ tensor in nD space.
- While ordinary Kronecker delta can be covariant, contravariant or mixed, the generalized Kronecker delta is always mixed with equal numbers of covariant and contravariant indices.

32. Give the inductive mathematical definition of the generalized Kronecker delta tensor $\delta^{i_1...i_n}_{j_1...j_n}$.
    Answer:
    $$\delta^{i_1...i_n}_{j_1...j_n} = \begin{cases} 1 & (j_1...j_n \text{ is even permutation of } i_1...i_n) \\ -1 & (j_1...j_n \text{ is odd permutation of } i_1...i_n) \\ 0 & (\text{repeated } i\text{'s or } j\text{'s}) \end{cases}$$

33. Write the determinantal array form of the generalized Kronecker delta tensor $\delta^{i_1...i_n}_{j_1...j_n}$ in terms of the ordinary Kronecker delta tensor.
    Answer:
    $$\delta^{i_1...i_n}_{j_1...j_n} = \begin{vmatrix} \delta^{i_1}_{j_1} & \delta^{i_1}_{j_2} & \cdots & \delta^{i_1}_{j_n} \\ \delta^{i_2}_{j_1} & \delta^{i_2}_{j_2} & \cdots & \delta^{i_2}_{j_n} \\ \vdots & \vdots & \ddots & \vdots \\ \delta^{i_n}_{j_1} & \delta^{i_n}_{j_2} & \cdots & \delta^{i_n}_{j_n} \end{vmatrix}$$

34. Define $\epsilon_{i_1...i_n}$ and $\epsilon^{i_1...i_n}$ in terms of the generalized Kronecker delta tensor.
    Answer:
    $$\epsilon_{i_1...i_n} = \delta^{1...n}_{i_1...i_n}$$
    $$\epsilon^{i_1...i_n} = \delta^{i_1...i_n}_{1...n}$$

35. Prove the relation: $\epsilon^{ijk}\epsilon_{lmn} = \delta^{ijk}_{lmn}$ using an analytic or an inductive or a truth table method.
    Answer: From the inductive definition of the entries of the permutation tensor we have:
    $$\epsilon^{ijk} = \begin{cases} 1 & (ijk \text{ is even permutation of 1,2,3}) \\ -1 & (ijk \text{ is odd permutation of 1,2,3}) \\ 0 & (\text{repeated index}) \end{cases}$$
    and
    $$\epsilon_{lmn} = \begin{cases} 1 & (lmn \text{ is even permutation of 1,2,3}) \\ -1 & (lmn \text{ is odd permutation of 1,2,3}) \\ 0 & (\text{repeated index}) \end{cases}$$

So, on multiplying ϵ^{ijk} with ϵ_{lmn} we have 9 cases: $\epsilon^{ijk}\epsilon_{lmn} = 1$ when ijk and lmn are of the same parity (i.e. both even permutations or both odd permutations) and $\epsilon^{ijk}\epsilon_{lmn} = -1$ when ijk and lmn are of different parity (i.e. one is even permutation and the other is odd permutation) while $\epsilon^{ijk}\epsilon_{lmn} = 0$ in all the remaining 5 cases (i.e. when one or both of ϵ^{ijk} and ϵ_{lmn} is zero due to repetition of index).

Now, if we look to the inductive definition of the generalized Kronecker delta tensor in a 3D space:

$$\delta^{ijk}_{lmn} = \begin{cases} 1 & (lmn \text{ is even permutation of } ijk) \\ -1 & (lmn \text{ is odd permutation of } ijk) \\ 0 & (\text{repeated index}) \end{cases}$$

we see it is essentially the same, i.e. $\delta^{ijk}_{lmn} = 1$ when ijk and lmn are of the same parity (in reference to 123), $\delta^{ijk}_{lmn} = -1$ when ijk and lmn are of opposite parity,[15] and $\delta^{ijk}_{lmn} = 0$ when the parity of one or both is ambiguous due to repetition of index. Hence, we conclude that $\epsilon^{ijk}\epsilon_{lmn}$ and δ^{ijk}_{lmn} are inductively equal and hence $\epsilon^{ijk}\epsilon_{lmn} = \delta^{ijk}_{lmn}$, as required.

36. Demonstrate that the generalized Kronecker delta is an absolute tensor.

    Answer: From the relation: $\delta^{i_1...i_n}_{j_1...j_n} = \epsilon^{i_1...i_n}\epsilon_{j_1...j_n}$ we see that the generalized Kronecker delta is equal to the product of the covariant permutation tensor (which is a relative tensor of weight $w = -1$ as proved earlier in Exercise 17) times the contravariant permutation tensor (which is a relative tensor of weight $w = +1$ as proved earlier in Exercise 17) and hence its weight is $w = 0$, i.e. it is an absolute tensor, as required. We note that the product of tensors is a tensor whose weight is the sum of the weights of the original tensors.

37. Prove the following relation justifying each step in your proof: $\delta^{mnq}_{klq} = \delta^{mn}_{kl}$.

    Answer:

    $$\begin{aligned}\delta^{mnq}_{klq} &= \begin{vmatrix} \delta^m_k & \delta^m_l & \delta^m_q \\ \delta^n_k & \delta^n_l & \delta^n_q \\ \delta^q_k & \delta^q_l & \delta^q_q \end{vmatrix} \\ &= \delta^m_k\left(\delta^n_l\delta^q_q - \delta^n_q\delta^q_l\right) - \delta^m_l\left(\delta^n_k\delta^q_q - \delta^n_q\delta^q_k\right) + \delta^m_q\left(\delta^n_k\delta^q_l - \delta^n_l\delta^q_k\right) \\ &= \delta^m_k\delta^n_l\delta^q_q - \delta^m_k\delta^n_q\delta^q_l - \delta^m_l\delta^n_k\delta^q_q + \delta^m_l\delta^n_q\delta^q_k + \delta^m_q\delta^n_k\delta^q_l - \delta^m_q\delta^n_l\delta^q_k \\ &= 3\delta^m_k\delta^n_l - \delta^m_k\delta^n_q\delta^q_l - 3\delta^m_l\delta^n_k + \delta^m_l\delta^n_q\delta^q_k + \delta^m_q\delta^n_k\delta^q_l - \delta^m_q\delta^n_l\delta^q_k \\ &= 3\delta^m_k\delta^n_l - \delta^m_k\delta^n_l - 3\delta^m_l\delta^n_k + \delta^m_l\delta^n_k + \delta^m_l\delta^n_k - \delta^m_k\delta^n_l \end{aligned}$$

---

[15] These implications are ultimately based on the rules of parity, that is:

$$\begin{aligned} \text{even} \times \text{even} &= \text{even} \\ \text{odd} \times \text{odd} &= \text{even} \\ \text{even} \times \text{odd} &= \text{odd} \\ \text{odd} \times \text{even} &= \text{odd} \end{aligned}$$

where the first implication is represented by the first two equalities while the second implication is represented by the last two equalities.

4 SPECIAL TENSORS

$$\begin{aligned}
&= \delta_k^m \delta_l^n - \delta_l^m \delta_k^n \\
&= \begin{vmatrix} \delta_k^m & \delta_l^m \\ \delta_k^n & \delta_l^n \end{vmatrix} \\
&= \delta_{kl}^{mn}
\end{aligned}$$

Justification:
Line 1 is based on the definition of the generalized Kronecker delta in 3D.
Line 2 is based on the definition of expansion of determinant.
Line 3 is based on the distributivity of multiplication over algebraic addition.
Line 4 is based on the identity $\delta_i^i = n$ with $n = 3$ (see Exercise 22).
Line 5 is based on using the ordinary Kronecker delta as an index replacement operator.
Line 6 is based on the operations of addition and subtraction.
Line 7 is based on the definition of expansion of determinant (in reverse).
Line 8 is based on the definition of the generalized Kronecker delta in 2D.

38. Prove the common form of the epsilon-delta identity.
    Answer: We have:

$$\begin{aligned}
\epsilon^{ijk}\epsilon_{lmk} &= \delta_{lmk}^{ijk} \\
&= \delta_{lm}^{ij} \\
&= \begin{vmatrix} \delta_l^i & \delta_m^i \\ \delta_l^j & \delta_m^j \end{vmatrix} \\
&= \delta_l^i \delta_m^j - \delta_m^i \delta_l^j
\end{aligned}$$

Justification:
Line 1 is based on the identity $\epsilon^{ijk}\epsilon_{lmn} = \delta_{lmn}^{ijk}$ (which we proved in Exercise 35) with $n = k$.
Line 2 is based on the identity $\delta_{klq}^{mnq} = \delta_{kl}^{mn}$ which we proved in the last question.
Line 3 is based on the definition of the generalized Kronecker delta in 2D.
Line 4 is based on the definition of expansion of determinant.

39. Prove the following generalization of the epsilon-delta identity:

$$g^{ij}\underline{\epsilon}_{ikl}\underline{\epsilon}_{jmn} = g_{km}g_{ln} - g_{kn}g_{lm}$$

Answer: On multiplying the two sides with $g^{km}g^{ln}$ we get:

$$g^{ij}g^{km}g^{ln}\underline{\epsilon}_{ikl}\underline{\epsilon}_{jmn} = g_{km}g_{ln}g^{km}g^{ln} - g_{kn}g_{lm}g^{km}g^{ln}$$

On raising the indices, we obtain:

$$\underline{\epsilon}^{jmn}\underline{\epsilon}_{jmn} = g_k^k g_l^l - g_n^m g_m^n$$

Now, from the identities: $\underline{\epsilon}^{jmn} = \frac{\epsilon^{jmn}}{\sqrt{g}}$, $\underline{\epsilon}_{jmn} = \sqrt{g}\epsilon_{jmn}$ and $g_j^i = \delta_j^i$ we get:

$$\epsilon^{jmn}\epsilon_{jmn} = \delta_k^k \delta_l^l - \delta_n^m \delta_m^n = \delta_k^k \delta_l^l - \delta_m^m$$

where the last step is based on employing the Kronecker delta as an index replacement operator. Noting that the rank and dimension of the permutation tensor are identical (i.e. 3 in this case), we have (using previously proved identities):[16]

$$\epsilon^{jmn}\epsilon_{jmn} = 3! = 6$$
$$\delta_k^k \delta_l^l - \delta_m^m = (3 \times 3) - 3 = 6$$

and hence the identity $\epsilon^{jmn}\epsilon_{jmn} = \delta_k^k \delta_l^l - \delta_m^m$ (which is obtained from the identity $g^{ij}\epsilon_{ikl}\epsilon_{jmn} = g_{km}g_{ln} - g_{kn}g_{lm}$) is correct. Therefore, the identity $g^{ij}\epsilon_{ikl}\epsilon_{jmn} = g_{km}g_{ln} - g_{kn}g_{lm}$ is valid.[17]

40. List and discuss all the main characteristics (e.g. symmetry) of the metric tensor.
Answer: The main characteristics are:
• It is a rank-2 symmetric absolute non-singular tensor.
• It has covariant, contravariant and mixed types with the latter being the same as the identity tensor.
• It is used for raising and lowering of indices and hence it facilitates the conversion of tensors from one variance type to another.
• It enters in the definition and formulation of many standard concepts and operations of tensor calculus and hence it is present everywhere in tensor calculus and its applications.
• It contains vital information about the main characteristic features of the space and its geometric properties.

41. How many types the metric tensor has?
Answer: It has three types: covariant g_{ij}, contravariant g^{ij} and mixed $g_i^j = \delta_i^j$.

42. Investigate the relation of the metric tensor of a given space to the coordinate systems of the space as well as its relation to the space itself by comparing the characteristics of the metric in different coordinate systems of the space such as being diagonal or not or having constant or variable components and so on. Hence, assess the status of the metric as a property of the space but with a form determined by the adopted coordinate system to describe the space and hence it is also a property of the coordinate system in this sense.
Answer: As a tensor, the metric should have significance regardless of any coordinate system where this significance is represented by the fact that it summarizes the characteristics of the space and its geometric nature such as being flat or curved. However, the metric tensor is closely related to the underlying coordinate system of the space through the relations:

$$g_{ij} = \mathbf{E}_i \cdot \mathbf{E}_j$$

---

[16] The relation $\delta_k^k \delta_l^l - \delta_m^m = (3 \times 3) - 3 = 6$ may not be obvious to some. Hence, we can relabel the dummy index m to obtain:

$$\delta_k^k \delta_l^l - \delta_m^m = \delta_k^k \delta_l^l - \delta_k^k = \delta_k^k \left(\delta_l^l - 1 \right) = 3 \times 2 = 3! = 6$$

[17] Alternatively, we may reverse the proof to obtain the required result.

4 SPECIAL TENSORS   67

$$g^{ij} = \mathbf{E}^i \cdot \mathbf{E}^j$$
$$g^i_j = \mathbf{E}^i \cdot \mathbf{E}_j$$

which link the entries of the metric tensor to the basis vectors which are intimately related to the coordinate system since the covariant basis vectors are the tangent vectors to the coordinate curves, while the contravariant basis vectors represent the gradient of the space coordinates and hence they are perpendicular to the coordinate surfaces. The close relation between the metric tensor and the coordinate system is reflected in many aspects. For example, the metric tensor is constant for rectilinear coordinate systems and variable for curvilinear coordinate systems because the basis vectors are constant for the former and variable for the latter. Another example is that the metric tensor is diagonal when the coordinate system of the space is orthogonal and this is justified by the above relations between the entries of the metric tensor and the basis vectors since the basis vectors of orthogonal systems are mutually perpendicular and hence the dot product will vanish when $i \neq j$. These examples, among others, represent the close relation between the adopted coordinate system and the form of the metric tensor.

Based on the above discussion, we can conclude that although the essence of the metric tensor is related to the nature of the space and hence it is independent of the coordinate system, the form of the metric tensor is highly dependent on the coordinate system. So in brief, the metric tensor summarizes essential properties of the space and hence it is independent of the employed coordinate system of the space, but the form of the metric tensor is highly dependent on the employed coordinate system and hence it is closely related to the employed coordinate system. Accordingly, the premise that is suggested in the question (i.e. the metric is a property of the space but with a form determined by the adopted coordinate system to describe the space and hence it is also a property of the coordinate system) is fully justified.

43. What is the relation between the covariant metric tensor and the length of an infinitesimal element of arc ds in a general coordinate system?
    Answer: The relation is:
    $$(ds)^2 = g_{ij} du^i du^j$$
    where g_{ij} is the covariant metric tensor, the indexed u are general coordinates and $i, j = 1 \cdots n$ with n being the dimension of the space.

44. How will the relation in question 43 become (a) in an orthogonal coordinate system and (b) in an orthonormal Cartesian coordinate system?
    Answer: In an orthogonal coordinate system it becomes:
    $$(ds)^2 = \sum_i (h_i)^2 du^i du^i$$
    where h_i is the scale factor of the ith coordinate and $i = 1 \cdots n$ with n being the dimension of the space.
    In an orthonormal Cartesian coordinate system it becomes:
    $$(ds)^2 = dx^i dx^i$$

where x^i is a Cartesian coordinate, $i = 1 \cdots n$ (with n being the dimension of the space) and sum over i is implied.

45. What is the characteristic feature of the metric tensor in orthogonal coordinate systems?
    Answer: It is diagonal, that is:

    $$\begin{aligned} g_{ij} &= 0 & g^{ij} &= 0 & (i \neq j) \\ g_{ij} &\neq 0 & g^{ij} &\neq 0 & (i = j) \end{aligned}$$

46. Write the mathematical expressions for the components of the covariant, contravariant and mixed forms of the metric tensor in terms of the covariant and contravariant basis vectors, $\mathbf{E}_i$ and $\mathbf{E}^i$.
    Answer:

    $$\begin{aligned} g_{ij} &= \mathbf{E}_i \cdot \mathbf{E}_j \\ g^{ij} &= \mathbf{E}^i \cdot \mathbf{E}^j \\ g^i_j &= \mathbf{E}^i \cdot \mathbf{E}_j \end{aligned}$$

47. Write, in full tensor notation, the mathematical expressions for the components of the covariant and contravariant forms of the metric tensor, g_{ij} and g^{ij}.
    Answer:

    $$\begin{aligned} g_{ij} &= \frac{\partial x^k}{\partial u^i} \frac{\partial x^k}{\partial u^j} \\ g^{ij} &= \frac{\partial u^i}{\partial x^k} \frac{\partial u^j}{\partial x^k} \end{aligned}$$

48. What is the relation between the mixed form of the metric tensor and the ordinary Kronecker delta tensor?
    Answer: They are identical, that is:

    $$g^i_j = \delta^i_j$$

49. Explain why the covariant and contravariant metric tensor is not necessarily diagonal in general coordinate systems but it is necessarily symmetric.
    Answer: The basis vectors, whether covariant or contravariant, in general coordinate systems are not necessarily mutually orthogonal and hence the metric tensor is not diagonal in general since the dot products which are given in the answer of a previous question (i.e. $g_{ij} = \mathbf{E}_i \cdot \mathbf{E}_j$ and $g^{ij} = \mathbf{E}^i \cdot \mathbf{E}^j$) are not necessarily zero when $i \neq j$. However, since the dot product of vectors is a commutative operation (i.e. $\mathbf{E}_i \cdot \mathbf{E}_j = \mathbf{E}_j \cdot \mathbf{E}_i$ and $\mathbf{E}^i \cdot \mathbf{E}^j = \mathbf{E}^j \cdot \mathbf{E}^i$), the metric tensor is necessarily symmetric (i.e. $g_{ij} = g_{ji}$ and $g^{ij} = g^{ji}$).

50. Explain why the diagonal elements of the covariant and contravariant metric tensor in general coordinate systems are not necessarily of unit magnitude or positive but they are necessarily non-zero.

Answer: Since the basis vectors, whether covariant or contravariant, in general coordinate systems are not necessarily of unit length, then the diagonal elements (which are given by $g_{ii} = \mathbf{E}_i \cdot \mathbf{E}_i = |\mathbf{E}_i|^2$ and $g^{ii} = \mathbf{E}^i \cdot \mathbf{E}^i = |\mathbf{E}^i|^2$ with no sum on i), are not necessarily of unit magnitude. Also, since the coordinates in some general coordinate systems can be imaginary, then the diagonal elements can be negative. However, because the basis vectors cannot vanish at the regular points of the space (since the tangent to the coordinate curve and the gradient to the coordinate surface do exist and they cannot be zero), then the dot products (i.e. $\mathbf{E}_i \cdot \mathbf{E}_i$ and $\mathbf{E}^i \cdot \mathbf{E}^i$), and hence the diagonal elements of the metric tensor (i.e. g_{ii} and g^{ii}), should necessarily be non-zero.

51. Explain why the mixed type metric tensor in any coordinate system is diagonal or in fact it is the unity tensor.
    Answer: As given earlier, the mixed type metric tensor is given by $g^i_j = \mathbf{E}^i \cdot \mathbf{E}_j$. Now, since the covariant and contravariant basis sets are reciprocal systems (i.e. the dot product between their corresponding elements is 1 and between their non-corresponding elements is 0), then we should have:

$$g^i_j = \mathbf{E}^i \cdot \mathbf{E}_j = 1 \qquad (i = j)$$
$$g^i_j = \mathbf{E}^i \cdot \mathbf{E}_j = 0 \qquad (i \neq j)$$

The second relation means that the mixed type metric tensor is diagonal and the first relation (combined with the second relation) means it is the unity tensor.
Alternatively, we have:

$$g_{ik} g^{kj} = \delta^j_i$$
$$g^j_i = \delta^j_i$$

where line 1 represents the fact that the covariant and contravariant metric tensors are inverses of each other while the second line is based on using the metric tensor as an index shifting operator. Hence, the mixed type metric tensor is diagonal and it is the unity tensor.

52. Show that the covariant and contravariant forms of the metric tensor, g_{ij} and g^{ij}, are inverses of each other.
    Answer: The covariant and contravariant forms of the metric tensor are given by:

$$g_{ij} = \frac{\partial x^k}{\partial u^i} \frac{\partial x^k}{\partial u^j}$$
$$g^{ij} = \frac{\partial u^i}{\partial x^k} \frac{\partial u^j}{\partial x^k}$$

Multiplying them as two matrices, we have:

$$g_{ij} g^{jk} = \frac{\partial x^m}{\partial u^i} \frac{\partial x^m}{\partial u^j} \frac{\partial u^j}{\partial x^n} \frac{\partial u^k}{\partial x^n}$$
$$= \frac{\partial x^m}{\partial u^i} \frac{\partial x^m}{\partial x^n} \frac{\partial u^k}{\partial x^n}$$

$$= \frac{\partial x^m}{\partial u^i} \frac{\partial x^n}{\partial x^m} \frac{\partial u^k}{\partial x^n}$$

$$= \frac{\partial x^m}{\partial u^i} \frac{\partial u^k}{\partial x^m}$$

$$= \frac{\partial u^k}{\partial u^i}$$

$$= \delta_i^k$$

where lines 2, 4 and 5 are the chain rule of differentiation, line 3 is the identity $\partial_i x_j = \partial_j x_i$ in orthonormal Cartesian systems (see Exercise 24), and line 6 is the identity $u^i_{,j} = \delta^i_j$ (see Exercise 23).

Similarly:

$$\begin{aligned}
g^{ij} g_{jk} &= \frac{\partial u^i}{\partial x^m} \frac{\partial u^j}{\partial x^m} \frac{\partial x^n}{\partial u^j} \frac{\partial x^n}{\partial u^k} \\
&= \frac{\partial u^i}{\partial x^m} \frac{\partial x^n}{\partial x^m} \frac{\partial x^n}{\partial u^k} \\
&= \frac{\partial u^i}{\partial x^m} \frac{\partial x^m}{\partial x^n} \frac{\partial x^n}{\partial u^k} \\
&= \frac{\partial u^i}{\partial x^m} \frac{\partial x^m}{\partial u^k} \\
&= \frac{\partial u^i}{\partial u^k} \\
&= \delta^i_k
\end{aligned}$$

where the lines are similarly justified as in the first part.

Hence, the covariant and contravariant forms of the metric tensor are inverses of each other, as required.

53. Why the determinant of the metric tensor should not vanish at any point in the space?
    Answer: Because the metric tensor should be invertible (i.e. non-singular) at every point in the space (i.e. we should have covariant and contravariant types globally if we are supposed to have an appropriate coordinate system).

54. If the determinant of the covariant metric tensor g_{ij} is g, what is the determinant of the contravariant metric tensor g^{ij}?
    Answer: It is the reciprocal of g, i.e. $1/g$. This can be easily inferred from the fact that the covariant and contravariant forms of the metric tensor are inverses of each other.

55. Show that the metric tensor can be regarded as a transformation of the ordinary Kronecker delta tensor in its different variance types from an orthonormal Cartesian coordinate system to a general coordinate system.
    Answer: From the definition of the metric tensor in its different variance types, we have:

$$g_{ij} = \frac{\partial x^k}{\partial u^i} \frac{\partial x^k}{\partial u^j}$$

4 SPECIAL TENSORS

$$g^{ij} = \frac{\partial u^i}{\partial x^k}\frac{\partial u^j}{\partial x^k}$$

$$g^i_j = \frac{\partial u^i}{\partial x^k}\frac{\partial x^k}{\partial u^j}$$

These equations can be written as:

$$g_{ij} = \frac{\partial x^k}{\partial u^i}\frac{\partial x^l}{\partial u^j}\delta_{kl}$$

$$g^{ij} = \frac{\partial u^i}{\partial x^k}\frac{\partial u^j}{\partial x^l}\delta^{kl}$$

$$g^i_j = \frac{\partial u^i}{\partial x^k}\frac{\partial x^l}{\partial u^j}\delta^k_l$$

where they are justified by the fact that the ordinary Kronecker delta tensor is an index replacement operator plus the fact that in orthonormal Cartesian systems all variance types of the Kronecker delta tensor are equal, i.e. $\delta_{ij}\mathbf{e}_i\mathbf{e}_j = \delta^{ij}\mathbf{e}_i\mathbf{e}_j = \delta^i_j\mathbf{e}_i\mathbf{e}_j$. As seen, the last three equations are the transformation equations of the ordinary Kronecker delta tensor in its different variance types from an orthonormal Cartesian coordinate system to a general coordinate system, as required.

56. Justify the use of the metric tensor as an index shifting operator using a mathematical argument.
Answer: Let have a vector $\mathbf{A}$ which can be covariant or contravariant and hence we have $\mathbf{A} = A_j\mathbf{E}^j = A^j\mathbf{E}_j$. So, let inner-multiply this vector with $\mathbf{E}_i$, that is:

$$\mathbf{A}\cdot\mathbf{E}_i = A_j\mathbf{E}^j\cdot\mathbf{E}_i = A_j\delta^j_i = A_i$$

$$\mathbf{A}\cdot\mathbf{E}_i = A^j\mathbf{E}_j\cdot\mathbf{E}_i = A^j g_{ji}$$

On comparing these two equations and taking account of the invariance of vector (i.e. its independence as a tensor from the employed basis set), we conclude that $A_i = A^j g_{ji}$, i.e. the covariant metric tensor is an index lowering operator.
Similarly, let inner-multiply this vector with $\mathbf{E}^i$, that is:

$$\mathbf{A}\cdot\mathbf{E}^i = A_j\mathbf{E}^j\cdot\mathbf{E}^i = A_j g^{ji}$$

$$\mathbf{A}\cdot\mathbf{E}^i = A^j\mathbf{E}_j\cdot\mathbf{E}^i = A^j\delta^i_j = A^i$$

On comparing these two equations and taking account of the invariance of vector, we conclude that $A^i = A_j g^{ji}$, i.e. the contravariant metric tensor is an index raising operator.
The above argument which employs a vector $\mathbf{A}$ can be easily generalized to non-scalar tensors of any rank and hence we conclude that the metric tensor is an index shifting operator, as required.

57. Carry out the following index shifting operations recording the order of the indices when necessary:

$$g^{ij}C_{klj} \qquad\qquad g_{mn}B^n_{\ st} \qquad\qquad g^l_n D_{km}{}^n$$

4 SPECIAL TENSORS 72

Answer:

$$\begin{aligned} g^{ij}C_{klj} &= C_{kl\cdot}{}^{i} \\ g_{mn}B^{n}{}_{\cdot st} &= B_{mst} \\ g^{l}_{n}D_{km\cdot}{}^{n} &= D_{km}{}^{l}. \end{aligned}$$

58. What is the difference between the three operations in question 57?
    Answer: The first is an index raising operation using the contravariant metric tensor, the second is an index lowering operation using the covariant metric tensor, and the third is an index replacement operation using the mixed type metric tensor (or Kronecker delta).

59. Why the order of the raised and lowered indices is important and hence it should be recorded? Mention one form of notation used to record the order of the indices.
    Answer: As indicated before, tensors are referred to basis vector set and hence the order of indices of tensors is important because it determines the order of the basis vectors to which the tensor is referred. We may also need to reverse the index shifting operation that we conducted earlier in a later stage. Therefore, the order of the raised and lowered indices is important and hence it should be kept and recorded. One form of notation that is used to record the order of the indices is to insert a dot to indicate the original position of the shifted index.

60. What is the condition that should be satisfied by the metric tensor of a flat space? Give common examples of flat and curved spaces.
    Answer: The metric tensor of a flat space can be represented by a diagonal form with all the diagonal elements being $+1$ or -1. Examples of flat spaces are plane, 3D Euclidean space and 4D Minkowski space. Examples of curved spaces are sphere and ellipsoid.

61. Considering a coordinate transformation, what is the relation between the determinants of the covariant metric tensor in the original and transformed coordinate systems, g and $\bar{g}$?
    Answer: The relation is:
    $$\bar{g} = J^2 g$$
    where J is the Jacobian of the transformation.

62. **B** is a "conjugate" or "associate" tensor of tensor **A**. What this means?
    Answer: It means that **B** is obtained from **A** by inner product multiplication of **A** by the covariant or contravariant metric tensor.

63. Complete and justify the following statement: "The components of the metric tensor are constants *iff* ...etc.".
    Answer: "The components of the metric tensor are constants *iff* the Christoffel symbols of the space metric vanish identically". This can be justified by the mathematical definitions of the Christoffel symbols of the first and second kind (which are given

4 SPECIAL TENSORS   73

in the text)[18] since the derivatives (and hence the Christoffel symbols) must vanish identically when the components of the metric tensor are constant. Similarly, when the Christoffel symbols vanish identically then in the general case the components of the metric tensor must be constant (refer to Exercise 20 in § 5 for more details).

64. What are the covariant and absolute derivatives of the metric tensor?
    Answer: The metric tensor is like a constant with respect to tensor differentiation and hence the covariant and absolute derivatives of the metric tensor are zero in all coordinate systems.

65. Assuming an orthogonal coordinate system of an nD space, complete the following equations where the indexed g represents the metric tensor or its components, $i \neq j$ in the second equation and there is no sum in the third equation:
$$\det\left(g^{ij}\right) =? \qquad g_{ij} =? \qquad g^{ii} =?$$
Answer:
$$\det\left(g^{ij}\right) = \frac{1}{g_{11}g_{22}\cdots g_{nn}}$$
$$g_{ij} = 0$$
$$g^{ii} = \frac{1}{g_{ii}}$$

66. Write, in matrix form, the covariant and contravariant metric tensor for orthonormal Cartesian, cylindrical and spherical coordinate systems. What distinguishes all these matrices? Explain and justify.
    Answer:
    Orthonormal Cartesian in 3D:
$$[g_{ij}] = [g^{ij}] = \begin{bmatrix} 1 & 0 & 0 \\ 0 & 1 & 0 \\ 0 & 0 & 1 \end{bmatrix}$$
Cylindrical:
$$[g_{ij}] = \begin{bmatrix} 1 & 0 & 0 \\ 0 & \rho^2 & 0 \\ 0 & 0 & 1 \end{bmatrix} \qquad [g^{ij}] = \begin{bmatrix} 1 & 0 & 0 \\ 0 & \frac{1}{\rho^2} & 0 \\ 0 & 0 & 1 \end{bmatrix}$$
Spherical:
$$[g_{ij}] = \begin{bmatrix} 1 & 0 & 0 \\ 0 & r^2 & 0 \\ 0 & 0 & r^2\sin^2\theta \end{bmatrix} \qquad [g^{ij}] = \begin{bmatrix} 1 & 0 & 0 \\ 0 & \frac{1}{r^2} & 0 \\ 0 & 0 & \frac{1}{r^2\sin^2\theta} \end{bmatrix}$$

---
[18] These definitions are:
$$[ij,k] = \frac{1}{2}\left(\partial_j g_{ik} + \partial_i g_{jk} - \partial_k g_{ij}\right)$$
$$\Gamma^k_{ij} = \frac{g^{kl}}{2}\left(\partial_j g_{il} + \partial_i g_{jl} - \partial_l g_{ij}\right)$$

All these matrices are diagonal because all these coordinate systems are orthogonal.

67. Referring to question 66, what is the relation between the diagonal elements of these matrices and the scale factors h_i of the coordinates of these systems?
Answer: The relation is:
$$g_{ii} = (h_i)^2 = \frac{1}{g^{ii}} \qquad \text{(no sum on } i\text{)}$$
where g_{ii} and g^{ii} are the ith diagonal elements of the covariant and contravariant metric tensor and h_i is the scale factor of the ith coordinate.

68. Considering the Minkowski metric, is the space of the mechanics of Lorentz transformations flat or curved? Is it homogeneous or not? What effect this can have on the length of element of arc ds?
Answer: The space of the mechanics of Lorentz transformations is flat because the metric tensor is diagonal with all the diagonal elements being $+1$ or -1, but it is not homogeneous because the metric tensor is not the unity tensor. The effect is that the length of element of arc ds can be imaginary.

69. Derive the following identities:
$$g_{im}\partial_k g^{mj} = -g^{mj}\partial_k g_{im} \qquad\qquad \partial_k g_{ij} = -g_{mj}g_{ni}\partial_k g^{nm}$$
Answer: We have $g_{im}g^{mj} = g_i^j = \delta_i^j$. On taking the partial derivative of both sides of this equation we obtain:
$$\begin{aligned}\partial_k\left(g_{im}g^{mj}\right) &= \partial_k \delta_i^j \\ g_{im}\partial_k g^{mj} + g^{mj}\partial_k g_{im} &= 0 \\ g_{im}\partial_k g^{mj} &= -g^{mj}\partial_k g_{im}\end{aligned}$$
which is the first relation. We note that the second line is justified by the product rule of differentiation and the fact that the components of the Kronecker delta tensor are constants.

Regarding the second relation, we start from the first relation and hence we have:
$$\begin{aligned}g_{im}\partial_k g^{mj} &= -g^{mj}\partial_k g_{im} \\ g^{mj}\partial_k g_{im} &= -g_{im}\partial_k g^{mj} \\ g^{nj}\partial_k g_{in} &= -g_{in}\partial_k g^{nj} \\ g_{mj}g^{nj}\partial_k g_{in} &= -g_{mj}g_{in}\partial_k g^{nj} \\ g_m^n \partial_k g_{in} &= -g_{mj}g_{in}\partial_k g^{nj} \\ \delta_m^n \partial_k g_{in} &= -g_{mj}g_{in}\partial_k g^{nj} \\ \partial_k g_{im} &= -g_{mj}g_{in}\partial_k g^{nj} \\ \partial_k g_{ij} &= -g_{jm}g_{in}\partial_k g^{nm} \\ \partial_k g_{ij} &= -g_{mj}g_{ni}\partial_k g^{nm}\end{aligned}$$
where in line 2 we exchange the two sides and multiply them by -1, in line 3 we relabel m as n, in line 4 we multiply both sides by g_{mj}, in line 5 we use the metric tensor as

an index shifting operator, in line 6 we use the identity $g_m^n = \delta_m^n$, in line 7 we use the Kronecker delta as an index replacement operator, in line 8 we exchange the labels of the indices m and j, and in line 9 we use the symmetry of the metric tensor.

70. What is the dot product of **A** and **B** where **A** is a rank-2 covariant tensor and **B** is a contravariant vector? Write this operation in steps providing full justification of each step.
Answer: We have $\mathbf{A} = A_{ij}\mathbf{E}^i\mathbf{E}^j$ and $\mathbf{B} = B^k\mathbf{E}_k$ and hence:

$$\begin{aligned}
\mathbf{A}\cdot\mathbf{B} &= \left(A_{ij}\mathbf{E}^i\mathbf{E}^j\right)\cdot\left(B^k\mathbf{E}_k\right) \\
&= A_{ij}B^k\left(\mathbf{E}^i\cdot\mathbf{E}_k\right)\mathbf{E}^j \\
&= A_{ij}B^k\delta^i_k\mathbf{E}^j \\
&= A_{ij}B^i\mathbf{E}^j
\end{aligned}$$

where line 1 is based on the definition of dot product and the definition of **A** and **B**, lines 2 and 3 are based on the definition of dot product of basis vectors and the relation between them and the Kronecker delta, and line 4 is based on using the Kronecker delta as an index replacement operator.
We can similarly have:

$$\begin{aligned}
\mathbf{A}\cdot\mathbf{B} &= \left(A_{ij}\mathbf{E}^i\mathbf{E}^j\right)\cdot\left(B^k\mathbf{E}_k\right) \\
&= A_{ij}B^k\left(\mathbf{E}^j\cdot\mathbf{E}_k\right)\mathbf{E}^i \\
&= A_{ij}B^k\delta^j_k\mathbf{E}^i \\
&= A_{ij}B^j\mathbf{E}^i
\end{aligned}$$

where the lines are similarly justified.

71. Derive an expression for the magnitude of a vector **A** when **A** is covariant and when **A** is contravariant.
Answer: If **A** is covariant then $\mathbf{A} = A_i\mathbf{E}^i$ and hence:

$$\begin{aligned}
|\mathbf{A}| &= \sqrt{\mathbf{A}\cdot\mathbf{A}} \\
&= \sqrt{(A_i\mathbf{E}^i)\cdot(A_j\mathbf{E}^j)} \\
&= \sqrt{(\mathbf{E}^i\cdot\mathbf{E}^j)A_iA_j} \\
&= \sqrt{g^{ij}A_iA_j} \\
&= \sqrt{A^jA_j}
\end{aligned}$$

where line 1 is a definition of the magnitude of a vector, line 2 is the definition of covariant vector, lines 3 and 4 are based on the definition of dot product of basis vectors and the relation between them and the metric tensor, and line 5 is based on using the metric tensor as an index shifting operator.
If **A** is contravariant then $\mathbf{A} = A^i\mathbf{E}_i$ and hence:

$$|\mathbf{A}| = \sqrt{\mathbf{A}\cdot\mathbf{A}}$$

4   SPECIAL TENSORS   76

$$\begin{aligned}
&= \sqrt{(A^i \mathbf{E}_i) \cdot (A^j \mathbf{E}_j)} \\
&= \sqrt{(\mathbf{E}_i \cdot \mathbf{E}_j) A^i A^j} \\
&= \sqrt{g_{ij} A^i A^j} \\
&= \sqrt{A_j A^j}
\end{aligned}$$

where these lines are similarly justified as in the first part.
As seen, these expressions are identical.

72. Derive an expression for the cosine of the angle θ between two covariant vectors, $\mathbf{A}$ and $\mathbf{B}$, and between two contravariant vectors $\mathbf{C}$ and $\mathbf{D}$.
    Answer: We have:

$$\begin{aligned}
\cos\theta &= \frac{\mathbf{A} \cdot \mathbf{B}}{|\mathbf{A}||\mathbf{B}|} \\
&= \frac{A_i \mathbf{E}^i \cdot B_j \mathbf{E}^j}{\sqrt{A_k \mathbf{E}^k \cdot A_l \mathbf{E}^l} \sqrt{B_m \mathbf{E}^m \cdot B_n \mathbf{E}^n}} \\
&= \frac{g^{ij} A_i B_j}{\sqrt{g^{kl} A_k A_l} \sqrt{g^{mn} B_m B_n}} \\
&= \frac{A^j B_j}{\sqrt{A^l A_l} \sqrt{B^n B_n}}
\end{aligned}$$

where line 1 is a definition of the cosine of angle between two vectors, line 2 is based on the definition of covariant vector and the definition of magnitude of a vector, line 3 is based on the relation between the dot product of basis vectors and the metric tensor, and line 4 is based on using the metric tensor as an index shifting operator.
Similarly, we have:

$$\begin{aligned}
\cos\theta &= \frac{\mathbf{C} \cdot \mathbf{D}}{|\mathbf{C}||\mathbf{D}|} \\
&= \frac{C^i \mathbf{E}_i \cdot D^j \mathbf{E}_j}{\sqrt{C^k \mathbf{E}_k \cdot C^l \mathbf{E}_l} \sqrt{D^m \mathbf{E}_m \cdot D^n \mathbf{E}_n}} \\
&= \frac{g_{ij} C^i D^j}{\sqrt{g_{kl} C^k C^l} \sqrt{g_{mn} D^m D^n}} \\
&= \frac{C_j D^j}{\sqrt{C_l C^l} \sqrt{D_n D^n}}
\end{aligned}$$

where these lines are similarly justified.
As seen, these expressions are identical (apart from the different labels of vectors).

73. What is the meaning of the angle between two intersecting smooth curves?
    Answer: It is the angle between the tangent vectors of these curves at the point of intersection.

4 SPECIAL TENSORS

74. What is the cross product of **A** and **B** where these are covariant vectors?
    Answer: It is:
    $$\mathbf{A} \times \mathbf{B} = \underline{\epsilon}^{ijk} A_i B_j \mathbf{E}_k$$
    where $\underline{\epsilon}^{ijk}$ is the absolute permutation tensor in its contravariant form.

75. Complete the following equations assuming a general coordinate system of a 3D space:
    $$\mathbf{E}_i \times \mathbf{E}_j = ? \qquad \mathbf{E}^i \times \mathbf{E}^j = ?$$
    Answer:
    $$\mathbf{E}_i \times \mathbf{E}_j = \underline{\epsilon}_{ijk} \mathbf{E}^k$$
    $$\mathbf{E}^i \times \mathbf{E}^j = \underline{\epsilon}^{ijk} \mathbf{E}_k$$
    where $\underline{\epsilon}_{ijk}$ and $\underline{\epsilon}^{ijk}$ are the absolute permutation tensor in its covariant and contravariant forms.

76. Define the operations of scalar triple product and vector triple product of vectors using tensor language and assuming a general coordinate system of a 3D space.
    Answer:
    The scalar triple product of covariant vectors **A**, **B** and **C** is given by:
    $$\mathbf{A} \cdot (\mathbf{B} \times \mathbf{C}) = \underline{\epsilon}^{ijk} A_i B_j C_k$$
    where $\underline{\epsilon}^{ijk}$ is the contravariant absolute permutation tensor.
    The scalar triple product of contravariant vectors **A**, **B** and **C** is given by:
    $$\mathbf{A} \cdot (\mathbf{B} \times \mathbf{C}) = \underline{\epsilon}_{ijk} A^i B^j C^k$$
    where $\underline{\epsilon}_{ijk}$ is the covariant absolute permutation tensor.
    The vector triple product of a covariant vector **A** and two contravariant vectors **B** and **C** is given by:
    $$\mathbf{A} \times (\mathbf{B} \times \mathbf{C}) = \epsilon^{ilm} \epsilon_{jkl} A_i B^j C^k \mathbf{E}_m$$
    The vector triple product of a contravariant vector **A** and two covariant vectors **B** and **C** is given by:
    $$\mathbf{A} \times (\mathbf{B} \times \mathbf{C}) = \epsilon_{ilm} \epsilon^{jkl} A^i B_j C_k \mathbf{E}^m$$

77. What is the relation between the relative and absolute permutation tensors in their covariant and contravariant forms?
    Answer: The covariant absolute permutation tensor $\underline{\epsilon}_{i_1 \ldots i_n}$ is equal to the covariant relative permutation tensor $\epsilon_{i_1 \ldots i_n}$ times $\sqrt{g}$, that is:
    $$\underline{\epsilon}_{i_1 \ldots i_n} = \sqrt{g}\, \epsilon_{i_1 \ldots i_n}$$
    where g is the determinant of the covariant metric tensor.
    The contravariant absolute permutation tensor $\underline{\epsilon}^{i_1 \ldots i_n}$ is equal to the contravariant relative permutation tensor $\epsilon^{i_1 \ldots i_n}$ divided by $\sqrt{g}$, that is:
    $$\underline{\epsilon}^{i_1 \ldots i_n} = \frac{1}{\sqrt{g}} \epsilon^{i_1 \ldots i_n}$$

78. Define the determinant of a matrix **B** in tensor notation assuming a general coordinate system of a 3D space.
    Answer: It is:
    $$\det(\mathbf{B}) = \frac{1}{3!} \delta^{ijk}_{lmn} B^l_i B^m_j B^n_k$$
    where δ^{ijk}_{lmn} is the generalized Kronecker delta of 3D space and **B** is represented by its mixed form.

79. Derive the relation for the length of line element in general coordinate systems: $(ds)^2 = g_{ij} du^i du^j$. How will this relation become when the coordinate system is orthogonal? Justify your answer.
    Answer: We have:
    $$\begin{aligned} (ds)^2 &= d\mathbf{r} \cdot d\mathbf{r} \\ &= \mathbf{E}_i du^i \cdot \mathbf{E}_j du^j \\ &= (\mathbf{E}_i \cdot \mathbf{E}_j) du^i du^j \\ &= g_{ij} du^i du^j \end{aligned}$$
    where **r** is the position vector in general coordinates, $\mathbf{E}_i$ and $\mathbf{E}_j$ are covariant basis vectors, u^i and u^j are general coordinates and g_{ij} is the covariant metric tensor. Line 1 is based on the definition of ds from first principles, line 2 is based on the definition of position vector in its infinitesimal differential form, line 3 is based on the definition of dot product, and line 4 is based on the relation between the dot product of basis vectors and the metric tensor.
    When the coordinate system is orthogonal, this relation becomes:
    $$(ds)^2 = \sum_i (h_i)^2 du^i du^i$$
    where h_i is the scale factor of the ith coordinate. The reason is that in orthogonal systems the metric tensor is diagonal where the ith diagonal (covariant) element is given by $g_{ii} = (h_i)^2$, and hence:
    $$(ds)^2 = g_{ij} du^i du^j = \sum_i (h_i)^2 du^i du^i$$

80. Write the integral representing the length L of a t-parameterized space curve in terms of the metric tensor.
    Answer: The integral is:
    $$L = \int_{t_1}^{t_2} \sqrt{g_{ij} \frac{du^i}{dt} \frac{du^j}{dt}} \, dt$$
    where t_1 and t_2 are the values of t corresponding to the start and end points of the curve respectively, g_{ij} is the covariant metric tensor of the space and u^i and u^j are general coordinates representing the path of the curve. As usual, we are assuming that the argument of the square root is non-negative since we are dealing with real values.

4 SPECIAL TENSORS

81. Using the equation $(ds)^2 = \sum_i (h_i)^2 du^i du^i$ plus the scale factors (which are given in a table in the book), develop expressions for ds in orthonormal Cartesian, cylindrical and spherical coordinate systems.
    Answer: In orthonormal Cartesian systems we have: $h_1 = h_2 = h_3 = 1$. Hence:
    $$\begin{aligned} (ds)^2 &= \sum_i (h_i)^2 du^i du^i \\ &= (dx^1)^2 + (dx^2)^2 + (dx^3)^2 \\ &= (dx)^2 + (dy)^2 + (dz)^2 \end{aligned}$$

    In cylindrical systems we have: $h_1 = h_3 = 1$ and $h_2 = \rho$. Hence:
    $$\begin{aligned} (ds)^2 &= \sum_i (h_i)^2 du^i du^i \\ &= (d\rho)^2 + \rho^2 (d\phi)^2 + (dz)^2 \end{aligned}$$

    In spherical systems we have: $h_1 = 1$, $h_2 = r$ and $h_3 = r \sin\theta$. Hence:
    $$\begin{aligned} (ds)^2 &= \sum_i (h_i)^2 du^i du^i \\ &= (dr)^2 + r^2 (d\theta)^2 + r^2 \sin^2\theta (d\phi)^2 \end{aligned}$$

82. Derive the following formula for the area of a differential element on the coordinate surface $u^i = $ constant in a 3D space assuming a general coordinate system:
    $$d\sigma(u^i = \text{constant}) = \sqrt{gg^{ii}} du^j du^k \qquad (i \neq j \neq k, \text{ no sum on } i)$$

    How will this relation become when the coordinate system is orthogonal?
    Answer: We have:
    $$\begin{aligned} d\sigma(u^i = C) &= |d\mathbf{r}_j \times d\mathbf{r}_k| \\ &= \left|\frac{\partial \mathbf{r}}{\partial u^j} \times \frac{\partial \mathbf{r}}{\partial u^k}\right| du^j du^k \\ &= |\mathbf{E}_j \times \mathbf{E}_k| du^j du^k \\ &= |\epsilon_{jki} \mathbf{E}^i| du^j du^k \\ &= |\epsilon_{jki}| |\mathbf{E}^i| du^j du^k \\ &= \sqrt{g}\sqrt{\mathbf{E}^i \cdot \mathbf{E}^i} du^j du^k \\ &= \sqrt{g}\sqrt{g^{ii}} du^j du^k \\ &= \sqrt{gg^{ii}} du^j du^k \end{aligned}$$

    where σ represents area, C is a constant, $d\mathbf{r}_j$ and $d\mathbf{r}_k$ are the infinitesimal displacement vectors along the jth and k^{th} coordinate curves that define the area element while the other symbols are as defined previously. We should also impose the condition $i \neq j \neq k$ with no sum over i in g^{ii}.

Line 1 is based on the definition of vector cross product and its relation to the area of the parallelogram that is defined by the vectors, line 2 is the chain rule in multi-variable differentiation, line 3 is based on the definition of covariant basis vectors as tangents to the coordinate curves, line 4 is based on the definition of cross product in general coordinate systems, line 5 is based on the rules of modulus, line 6 is based on the relation $\underline{\epsilon}_{jki} = \sqrt{g}\epsilon_{jki}$ and the fact that $|\epsilon_{jki}| = 1$ (since $i \neq j \neq k$) plus the definition of the magnitude of a vector, and line 7 is based on the relation between the contravariant basis vectors and the elements of the contravariant metric tensor.

In orthogonal coordinate systems of a 3D space we have:

$$\sqrt{g g^{ii}} = \sqrt{(h_i)^2 (h_j)^2 (h_k)^2 \frac{1}{(h_i)^2}} = h_j h_k \qquad (i \neq j \neq k, \text{ no sum on any index})$$

and hence the above relation becomes:

$$d\sigma(u^i = C) = h_j h_k du^j du^k$$

where $i \neq j \neq k$ and no sum on j or k.

83. Using the equation $d\sigma(u^i = C) = h_j h_k du^j du^k$ plus the scale factors (which are given in a table in the book), develop expressions for $d\sigma$ on the coordinate surfaces in orthonormal Cartesian, cylindrical and spherical coordinate systems.
Answer: In orthonormal Cartesian systems we have: $h_1 = h_2 = h_3 = 1$. Hence:

$$\begin{aligned} d\sigma(x = C) &= dydz \\ d\sigma(y = C) &= dxdz \\ d\sigma(z = C) &= dxdy \end{aligned}$$

In cylindrical systems we have: $h_1 = h_3 = 1$ and $h_2 = \rho$. Hence:

$$\begin{aligned} d\sigma(\rho = C) &= \rho d\phi dz \\ d\sigma(\phi = C) &= d\rho dz \\ d\sigma(z = C) &= \rho d\rho d\phi \end{aligned}$$

In spherical systems we have: $h_1 = 1$, $h_2 = r$ and $h_3 = r\sin\theta$. Hence:

$$\begin{aligned} d\sigma(r = C) &= r^2 \sin\theta d\theta d\phi \\ d\sigma(\theta = C) &= r \sin\theta dr d\phi \\ d\sigma(\phi = C) &= r dr d\theta \end{aligned}$$

84. Derive the following formula for the volume of a differential element of a solid body in a 3D space assuming a general coordinate system:

$$d\tau = \sqrt{g}\, du^1 du^2 du^3$$

How will this relation become when the coordinate system is orthogonal?
Answer: We have:

$$\begin{aligned}
d\tau &= |d\mathbf{r}_1 \cdot (d\mathbf{r}_2 \times d\mathbf{r}_3)| \\
&= \left|\frac{\partial \mathbf{r}}{\partial u^1} \cdot \left(\frac{\partial \mathbf{r}}{\partial u^2} \times \frac{\partial \mathbf{r}}{\partial u^3}\right)\right| du^1 du^2 du^3 \\
&= |\mathbf{E}_1 \cdot (\mathbf{E}_2 \times \mathbf{E}_3)| \, du^1 du^2 du^3 \\
&= \left|\mathbf{E}_1 \cdot \underline{\epsilon}_{231}\mathbf{E}^1\right| du^1 du^2 du^3 \\
&= \left|\mathbf{E}_1 \cdot \mathbf{E}^1\right| |\underline{\epsilon}_{231}| \, du^1 du^2 du^3 \\
&= |\delta^1_1| \, |\underline{\epsilon}_{231}| \, du^1 du^2 du^3 \\
&= \sqrt{g} \, du^1 du^2 du^3
\end{aligned}$$

where τ represents volume, $d\mathbf{r}_1, d\mathbf{r}_2, d\mathbf{r}_3$ are the three infinitesimal displacement vectors that define the element of volume along the corresponding coordinate curves, g is the determinant of the covariant metric tensor g_{ij} and the other symbols are as defined previously.

Line 1 is based on the definition of scalar triple product of vectors and its relation to the volume of the parallelepiped that is defined by the vectors, line 2 is the chain rule in multi-variable differentiation, line 3 is based on the definition of covariant basis vectors as tangents to the coordinate curves, line 4 is based on the definition of cross product in general coordinate systems, line 5 is based on the rules of modulus, line 6 is based on the relation between the basis vectors and the Kronecker delta, and line 7 is based on the relation $\underline{\epsilon}_{231} = \sqrt{g}\epsilon_{231}$ and the fact that $|\epsilon_{231}| = 1$ plus the fact that $|\delta^1_1| = 1$.

In orthogonal coordinate systems of a 3D space we have $\sqrt{g} = h_1 h_2 h_3$ and hence the above relation becomes:

$$d\tau = h_1 h_2 h_3 \, du^1 du^2 du^3$$

85. Make a plot representing the volume of an infinitesimal element of a solid body in a 3D space as the magnitude of a scalar triple product of three vectors.
    Answer: The plot should look similar to Figure 7.

86. Use the expression of the volume element in general coordinate systems of nD spaces to find the formula for the volume element in orthogonal coordinate systems.
    Answer: The generalized volume element $d\tau$ in general coordinate systems of nD spaces is given by the formula:

    $$d\tau = \sqrt{g} du^1 \ldots du^n$$

    where g is the determinant of the covariant metric tensor g_{ij} and the indexed u are general coordinates. In orthogonal coordinate systems we have $\sqrt{g} = h_1 \cdots h_n$ and hence the formula becomes:

    $$d\tau = h_1 \cdots h_n du^1 \ldots du^n$$

    with no summation over n.

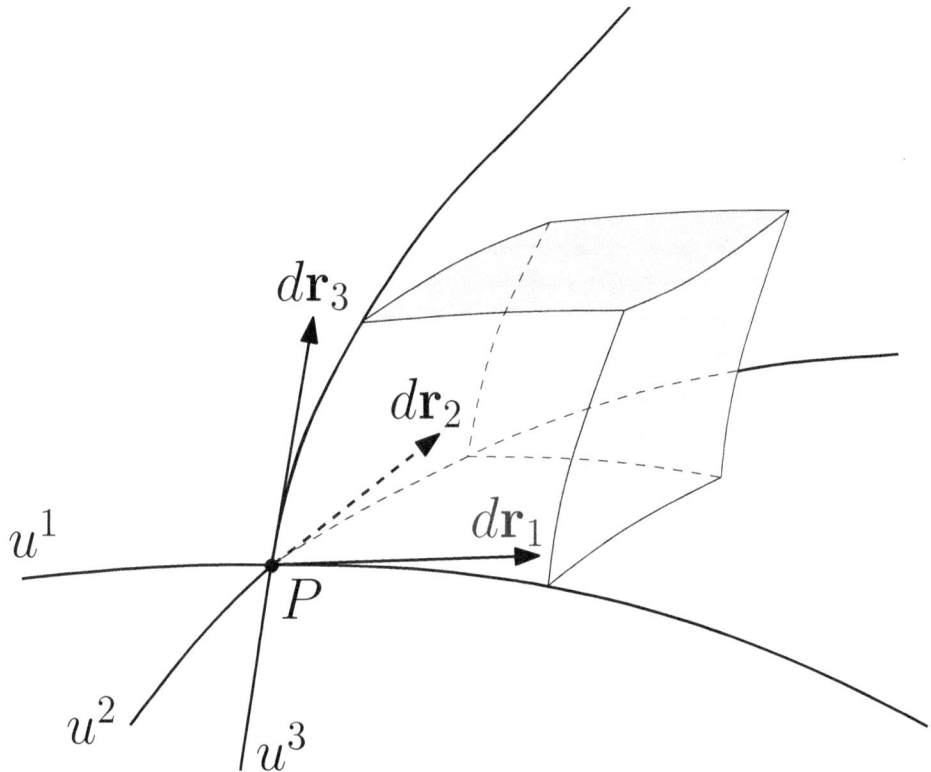

Figure 7: The volume of an infinitesimal element of a solid body in a 3D space in the neighborhood of a given point P as the magnitude of the scalar triple product of the infinitesimal displacement vectors in the directions of the three coordinate curves at P, $d\mathbf{r}_1$, $d\mathbf{r}_2$ and $d\mathbf{r}_3$.

87. Using the equation $d\tau = h_1 h_2 h_3 \, du^1 du^2 du^3$ plus the scale factors (which are given in a table in the book), develop expressions for $d\tau$ in orthonormal Cartesian, cylindrical and spherical coordinate systems.
    Answer: In orthonormal Cartesian systems we have: $h_1 = h_2 = h_3 = 1$. Hence:
    $$d\tau = dxdydz$$
    In cylindrical systems we have: $h_1 = h_3 = 1$ and $h_2 = \rho$. Hence:
    $$d\tau = \rho d\rho d\phi dz$$
    In spherical systems we have: $h_1 = 1$, $h_2 = r$ and $h_3 = r\sin\theta$. Hence:
    $$d\tau = r^2 \sin\theta dr d\theta d\phi$$

Chapter 5
Tensor Differentiation

1. Why tensor differentiation (represented by covariant and absolute derivatives) is needed in general coordinate systems to replace the ordinary differentiation (represented by partial and total derivatives)?
   Answer: Because the basis vectors in the general coordinate systems are coordinate dependent and hence the differentiation of tensors should extend to the basis vectors and not restricted to the components and this results in the tensor differentiation, i.e. covariant and absolute differentiation. In brief, tensor differentiation is the differentiation of a tensor in its two parts (i.e. the components and the basis vectors to which these components are referred) since both these parts are coordinate dependent and hence they are variable and should be subject to the process of differentiation.

2. Show that in general coordinate systems, the ordinary differentiation of the components of non-scalar tensors with respect to the coordinates will not produce a tensor in general.
   Answer: Let have a tensor $\mathbf{A}$ of type (m, n) coordinated by a general coordinate system and hence $\mathbf{A} = A^{i_1,\cdots,i_m}_{j_1,\cdots,j_n} \mathbf{E}_{i_1} \cdots \mathbf{E}_{i_m} \mathbf{E}^{j_1} \cdots \mathbf{E}^{j_n}$ where both $A^{i_1,\cdots,i_m}_{j_1,\cdots,j_n}$ and $\mathbf{E}_{i_1} \cdots \mathbf{E}_{i_m} \mathbf{E}^{j_1} \cdots \mathbf{E}^{j_n}$ are variables that depend on coordinates. Now, let differentiate this tensor with respect to the kth coordinate using the product rule of differentiation, that is:

$$\partial_k \mathbf{A} = \left(\partial_k A^{i_1,\cdots,i_m}_{j_1,\cdots,j_n}\right) \mathbf{E}_{i_1} \cdots \mathbf{E}_{i_m} \mathbf{E}^{j_1} \cdots \mathbf{E}^{j_n} + A^{i_1,\cdots,i_m}_{j_1,\cdots,j_n} \partial_k \left(\mathbf{E}_{i_1} \cdots \mathbf{E}_{i_m} \mathbf{E}^{j_1} \cdots \mathbf{E}^{j_n}\right)$$

As we see, the ordinary differentiation is valid only if the second term on the right hand side vanishes identically which is not the case in general coordinate systems since the basis vectors in these systems are not constant. Accordingly, the ordinary differentiation is just part of tensor differentiation and hence it will not produce a tensor in general since a tensor is produced only by the full differentiation process of a tensor.

3. "The Christoffel symbols are affine tensors but not tensors". Explain and justify this statement.
   Answer: It means that these symbols transform like tensors in affine coordinate systems but not in general coordinate systems. The justification is that: let transform a Christoffel symbol of the first kind from one general system (unbarred) to another general system (barred) to see if it will transform like a tensor or not. To ease the notation, let use lower case indices for the original system (unbarred) and upper case indices for the transformed system (barred). Accordingly, the covariant metric tensor is transformed by the following equation:

$$g_{IJ} = g_{rs} \frac{\partial u^r}{\partial u^I} \frac{\partial u^s}{\partial u^J} = g_{rs} \partial_I u^r \partial_J u^s \tag{1}$$

5 TENSOR DIFFERENTIATION

The Christoffel symbols of the first kind in the barred system are defined as:

$$[IJ, K] = \frac{1}{2} \left(\partial_J g_{KI} + \partial_I g_{JK} - \partial_K g_{IJ} \right) \qquad (2)$$

So, let take the partial derivative of both sides of Eq. 1 with respect to index K and substitute in Eq. 2, that is:

$$\begin{aligned} \partial_K g_{IJ} &= \partial_K \left(g_{rs} \partial_I u^r \partial_J u^s \right) \\ &= \partial_K g_{rs} \partial_I u^r \partial_J u^s + g_{rs} \partial_{IK} u^r \partial_J u^s + g_{rs} \partial_I u^r \partial_{JK} u^s \\ &= \partial_t g_{rs} \partial_K u^t \partial_I u^r \partial_J u^s + g_{rs} \partial_{IK} u^r \partial_J u^s + g_{rs} \partial_I u^r \partial_{JK} u^s \end{aligned}$$

where the last step is justified by the chain rule. By using cyclic replacement of indices in the last equation, we obtain:

$$\begin{aligned} \partial_J g_{KI} &= \partial_s g_{tr} \partial_J u^s \partial_K u^t \partial_I u^r + g_{tr} \partial_{KJ} u^t \partial_I u^r + g_{tr} \partial_K u^t \partial_{IJ} u^r \\ \partial_I g_{JK} &= \partial_r g_{st} \partial_I u^r \partial_J u^s \partial_K u^t + g_{st} \partial_{JI} u^s \partial_K u^t + g_{st} \partial_J u^s \partial_{KI} u^t \end{aligned}$$

On relabeling the dummy indices to unify the notation and noting the commutativity of partial differential operators (i.e. $\partial_i \partial_j = \partial_j \partial_i$), we obtain the following relations:

$$\begin{aligned} \partial_K g_{IJ} &= \partial_t g_{rs} \partial_K u^t \partial_I u^r \partial_J u^s + g_{rs} \partial_{IK} u^r \partial_J u^s + g_{rs} \partial_I u^r \partial_{JK} u^s \\ \partial_J g_{KI} &= \partial_t g_{rs} \partial_J u^t \partial_K u^r \partial_I u^s + g_{rs} \partial_{JK} u^r \partial_I u^s + g_{rs} \partial_K u^r \partial_{IJ} u^s \\ \partial_I g_{JK} &= \partial_t g_{rs} \partial_I u^t \partial_J u^r \partial_K u^s + g_{rs} \partial_{IJ} u^r \partial_K u^s + g_{rs} \partial_J u^r \partial_{IK} u^s \end{aligned}$$

On substituting from these expressions into Eq. 2 (marking canceled terms with similar type of brackets and noting the symmetry of the metric tensor and the insignificance of the labels of dummy indices) we obtain:

$$\begin{aligned} 2[IJ, K] &= \partial_J g_{KI} + \partial_I g_{JK} - \partial_K g_{IJ} \\ &= \partial_t g_{rs} \partial_J u^t \partial_K u^r \partial_I u^s + (g_{rs} \partial_{JK} u^r \partial_I u^s) + g_{rs} \partial_K u^r \partial_{IJ} u^s + \\ &\quad \partial_t g_{rs} \partial_I u^t \partial_J u^r \partial_K u^s + g_{rs} \partial_{IJ} u^r \partial_K u^s + [g_{rs} \partial_J u^r \partial_{IK} u^s] - \\ &\quad \partial_t g_{rs} \partial_K u^t \partial_I u^r \partial_J u^s - [g_{rs} \partial_{IK} u^r \partial_J u^s] - (g_{rs} \partial_I u^r \partial_{JK} u^s) \\ &= \partial_t g_{rs} \partial_J u^t \partial_K u^r \partial_I u^s + \partial_t g_{rs} \partial_I u^t \partial_J u^r \partial_K u^s - \partial_t g_{rs} \partial_K u^t \partial_I u^r \partial_J u^s \\ &\quad + g_{rs} \partial_K u^r \partial_{IJ} u^s + g_{rs} \partial_{IJ} u^r \partial_K u^s \end{aligned}$$

On relabeling the dummy indices in the last equation noting the symmetry of the metric tensor we get:

$$\begin{aligned} 2[IJ, K] &= \partial_t g_{rs} \partial_J u^t \partial_K u^r \partial_I u^s + \partial_s g_{tr} \partial_I u^s \partial_J u^t \partial_K u^r - \partial_r g_{st} \partial_K u^r \partial_I u^s \partial_J u^t \\ &\quad + g_{rs} \partial_K u^r \partial_{IJ} u^s + g_{rs} \partial_{IJ} u^s \partial_K u^r \\ &= \left(\partial_t g_{rs} + \partial_s g_{tr} - \partial_r g_{st} \right) \partial_J u^t \partial_K u^r \partial_I u^s \\ &\quad + 2 g_{rs} \partial_K u^r \partial_{IJ} u^s \end{aligned}$$

5 TENSOR DIFFERENTIATION

$$= 2\left[st,r\right]\partial_J u^t \partial_K u^r \partial_I u^s + 2g_{rs}\partial_K u^r \partial_{IJ}u^s$$

On dividing both sides by 2 we obtain:

$$\begin{aligned}[IJ,K] &= [st,r]\partial_J u^t \partial_K u^r \partial_I u^s + g_{rs}\partial_K u^r \partial_{IJ}u^s \\ &= [st,r]\frac{\partial u^t}{\partial u^J}\frac{\partial u^r}{\partial u^K}\frac{\partial u^s}{\partial u^I} + g_{rs}\frac{\partial u^r}{\partial u^K}\frac{\partial^2 u^s}{\partial u^I \partial u^J}\end{aligned}$$

i.e. the Christoffel symbol of the first kind transforms like a tensor but with an added term (i.e. the second term on the right hand side of the last equation). Hence, the Christoffel symbol of the first kind transforms like a tensor only if the second term vanishes identically and this is true only in affine systems where the second partial derivative is zero, i.e. the Christoffel symbols of the first kind are affine tensors but not general tensors.

The above argument can be easily extended to the Christoffel symbols of the second kind since they are obtained from the Christoffel symbols of the first kind by raising an index and hence if the original "tensor" is only an affine tensor and not a general tensor then the raised "tensor" should also be an affine tensor and not a general tensor because the operation of raising indices does not change the nature of tensor.

4. What is the difference between the first and second kinds of the Christoffel symbols?
   Answer: They differ in an index where it is covariant in the first kind and contravariant in the second kind.

5. Show that the Christoffel symbols of both kinds are not general tensors by giving examples of these symbols being vanishing in some systems but not in other systems and considering the universality of the zero tensor.
   Answer: As we established earlier, the zero tensor is invariant across all coordinate systems and hence if a tensor is zero in one system it must be zero in all other systems. Accordingly, if the Christoffel symbols of both kinds are tensors then when they vanish in one coordinate system they must vanish in all other systems. However, this is not true since the Christoffel symbols of both kinds vanish in Cartesian coordinate systems but not in cylindrical or spherical coordinate systems for instance and hence they cannot be tensors unconditionally, i.e. they are not general tensors. However, since they vanish across all affine systems (and hence they are invariant across all these systems) then they should be affine tensors.

6. State the mathematical definitions of the Christoffel symbols of the first and second kinds. How these two kinds are obtained from each other?
   Answer: The Christoffel symbols of the first and second kind are defined respectively by:

$$\begin{aligned}[ij,k] &= \frac{1}{2}\left(\partial_j g_{ik} + \partial_i g_{jk} - \partial_k g_{ij}\right) \\ \Gamma^k_{ij} &= \frac{g^{kl}}{2}\left(\partial_j g_{il} + \partial_i g_{jl} - \partial_l g_{ij}\right)\end{aligned}$$

where the indexed g are the covariant and contravariant forms of the metric tensor. As seen, the second kind is obtained from the first kind by raising an index while the first kind is obtained from the second kind by lowering an index.

7. What is the significance of the Christoffel symbols being solely dependent on the coefficients of the metric tensor in their relation to the underlying space and coordinate system?
   Answer: The sole dependency of the Christoffel symbols on the coefficients of the metric tensor means that they are variable functions of the coordinates of the space like the coefficients of the metric tensor. Hence, when the coefficients are constants (i.e. independent of coordinates) the Christoffel symbols will vanish identically over the whole space. In this case, the Christoffel symbols will reflect a property of the space (i.e. being flat) and a property of the system (i.e. being affine or rectilinear).
8. Do the Christoffel symbols represent a property of the space, a property of the coordinate system, or a property of both?
   Answer: As discussed earlier, the metric tensor contains essential information about the geometric nature of the space and hence it is a property of the space and depends on the nature of the space. However, the metric tensor also depends in form on the employed system for coordinating the space and hence it can also be regarded as a property of the coordinate system. Accordingly, the dependency of the Christoffel symbols on the coefficients of the metric tensor means that they are a characteristic property of both the underlying space and the employed coordinate system and hence they reflect certain features of the geometric nature of the space as well as certain features of the employed coordinate system.
9. If some of the Christoffel symbols vanish in a particular curvilinear coordinate system, should these some necessarily vanish in other curvilinear coordinate systems? Justify your answer by giving some examples.
   Answer: No. For example, in cylindrical systems we have only 3 non-vanishing Christoffel symbols but in spherical systems we have 9 non-vanishing symbols and hence some of the vanishing symbols in cylindrical systems, such as [33, 1], correspond to non-vanishing symbols in spherical systems.
10. Verify that the Christoffel symbols of the first and second kind are symmetric in their paired indices by using their mathematical definitions.
    Answer: The Christoffel symbols of the first kind are given by:
    $$[ij,l] = \frac{1}{2}\left(\partial_j g_{il} + \partial_i g_{jl} - \partial_l g_{ij}\right)$$
    On shifting the paired indices we obtain:
    $$\begin{aligned} [ji,l] &= \frac{1}{2}\left(\partial_i g_{jl} + \partial_j g_{il} - \partial_l g_{ji}\right) \\ &= \frac{1}{2}\left(\partial_j g_{il} + \partial_i g_{jl} - \partial_l g_{ji}\right) \\ &= \frac{1}{2}\left(\partial_j g_{il} + \partial_i g_{jl} - \partial_l g_{ij}\right) \\ &= [ij,l] \end{aligned}$$
    where in line 2 we just exchanged the first and second terms, while line 3 is justified by the symmetry of the metric tensor.

5   *TENSOR DIFFERENTIATION*   87

Similarly, the Christoffel symbols of the second kind are given by:

$$\Gamma^k_{ij} = \frac{g^{kl}}{2}\left(\partial_j g_{il} + \partial_i g_{jl} - \partial_l g_{ij}\right)$$

On shifting the paired indices we obtain:

$$\begin{aligned}
\Gamma^k_{ji} &= \frac{g^{kl}}{2}\left(\partial_i g_{jl} + \partial_j g_{il} - \partial_l g_{ji}\right) \\
&= \frac{g^{kl}}{2}\left(\partial_j g_{il} + \partial_i g_{jl} - \partial_l g_{ji}\right) \\
&= \frac{g^{kl}}{2}\left(\partial_j g_{il} + \partial_i g_{jl} - \partial_l g_{ij}\right) \\
&= \Gamma^k_{ij}
\end{aligned}$$

where in line 2 we just exchanged the first and second terms, while line 3 is justified by the symmetry of the metric tensor.

11. Correct, if necessary, the following equations:

$$\partial_j \mathbf{E}_i = -\Gamma^k_{ij}\mathbf{E}_k \qquad\qquad \partial_j \mathbf{E}^i = -\Gamma^i_{mj}\mathbf{E}^m$$

Answer: The second equation is correct while the first equation should be corrected by removing the minus sign.

12. What is the significance of the following equations?

$$\mathbf{E}^k \cdot \partial_j \mathbf{E}_i = \Gamma^k_{ij} \qquad\qquad \mathbf{E}_k \cdot \partial_j \mathbf{E}^i = -\Gamma^i_{kj} \qquad\qquad \mathbf{E}_k \cdot \partial_j \mathbf{E}_i = [ij,k]$$

Answer: The first and second equations mean that the Christoffel symbols of the second kind are the projections of the partial derivative of the basis vectors in the direction of the basis vectors of the opposite variance type, while the third equation means that the Christoffel symbols of the first kind are the projections of the partial derivative of the covariant basis vectors in the direction of the basis vectors of the same variance type.

13. Derive the following relations giving full explanation of each step:

$$\partial_j g_{il} = [ij,l] + [lj,i] \qquad\qquad \Gamma^j_{ji} = \partial_i\left(\ln\sqrt{g}\right)$$

Answer: Regarding the first relation, from the definition of the Christoffel symbols of the first kind we have:

$$\begin{aligned}
[ij,l] &= \frac{1}{2}\left(\partial_j g_{il} + \partial_i g_{jl} - \partial_l g_{ij}\right) \\
[lj,i] &= \frac{1}{2}\left(\partial_j g_{li} + \partial_l g_{ji} - \partial_i g_{lj}\right)
\end{aligned}$$

where the second equation is obtained from the first by exchanging i and l. On adding the two sides of these equations we obtain:

$$[ij,l] + [lj,i] = \frac{1}{2}\left(\partial_j g_{il} + \partial_i g_{jl} - \partial_l g_{ij} + \partial_j g_{li} + \partial_l g_{ji} - \partial_i g_{lj}\right)$$

$$= \frac{1}{2}\left(\partial_j g_{il} + \partial_i g_{jl} - \partial_l g_{ij} + \partial_j g_{il} + \partial_l g_{ij} - \partial_i g_{jl}\right)$$
$$= \frac{1}{2}\left(\partial_j g_{il} + \partial_j g_{il}\right)$$
$$= \partial_j g_{il}$$

which is the required result. We note that line 2 is justified by the symmetry of the metric tensor.

Regarding the second relation, the Christoffel symbols of the second kind are given by:

$$\Gamma^k_{ij} = \Gamma^k_{ji} = \frac{g^{kl}}{2}\left(\partial_j g_{il} + \partial_i g_{jl} - \partial_l g_{ij}\right)$$

where the symmetry in i and j (according to the first equality) was justified in Exercise 10. On contracting k with j we obtain:

$$\begin{aligned}
\Gamma^j_{ji} &= \frac{g^{jl}}{2}\left(\partial_j g_{il} + \partial_i g_{jl} - \partial_l g_{ij}\right) \\
&= \frac{1}{2}\left(g^{jl}\partial_j g_{il} + g^{jl}\partial_i g_{jl} - g^{jl}\partial_l g_{ij}\right) \\
&= \frac{1}{2}\left(g^{lj}\partial_l g_{ij} + g^{jl}\partial_i g_{jl} - g^{jl}\partial_l g_{ij}\right) \\
&= \frac{1}{2}\left(g^{jl}\partial_l g_{ij} + g^{jl}\partial_i g_{jl} - g^{jl}\partial_l g_{ij}\right) \\
&= \frac{1}{2}g^{jl}\partial_i g_{jl} \\
&= \frac{1}{2g}gg^{jl}\partial_i g_{jl} \\
&= \frac{1}{2g}\partial_i g \\
&= \frac{1}{2}\partial_i\left(\ln g\right) \\
&= \partial_i\left(\frac{1}{2}\ln g\right) \\
&= \partial_i\left(\ln\sqrt{g}\right)
\end{aligned}$$

which is the required result. We note that in line 3 we relabeled the dummy indices j and l in the first term, in line 4 we used the symmetry of the metric tensor (i.e. $g^{lj} = g^{jl}$), in line 7 we used the expression of the derivative of the determinant g of the covariant metric tensor g_{ij} (see Exercise 17), in line 8 we used the rule of differentiation of natural logarithm, and in line 10 we used the rule of power of logarithm.

14. Assuming an orthogonal coordinate system, verify the following relation: $[ij,k] = 0$ where $i \neq j \neq k$.

    Answer: In orthogonal coordinate systems the metric tensor is diagonal and hence

$g_{mn} = 0$ when $m \neq n$. From the definition of the Christoffel symbols of the first kind we have:
$$[ij, k] = \frac{1}{2} \left(\partial_j g_{ik} + \partial_i g_{jk} - \partial_k g_{ij} \right)$$
Therefore, when $i \neq j \neq k$ then we have $g_{ik} = g_{jk} = g_{ij} = 0$ and hence $[ij, k] = 0$.

15. Assuming an orthogonal coordinate system, verify the following relation: $\Gamma^i_{ji} = \frac{1}{2} \partial_j \ln g_{ii}$ with no sum over i.
    Answer: According to the definition of the Christoffel symbols of the second kind we have:
    $$\Gamma^i_{jk} = g^{il} [jk, l]$$
    In orthogonal coordinate systems the metric tensor is diagonal and hence $g^{il} = 0$ when $i \neq l$. Therefore, the above equation becomes:
    $$\Gamma^i_{jk} = g^{ii} [jk, i]$$
    with no sum over i. Moreover, in orthogonal systems we have:
    $$g^{ii} = \frac{1}{g_{ii}} \qquad \text{(no sum over } i\text{)}$$
    and hence:
    $$\Gamma^i_{jk} = \frac{[jk, i]}{g_{ii}} = \frac{1}{2g_{ii}} \left(\partial_k g_{ij} + \partial_j g_{ki} - \partial_i g_{jk} \right)$$
    where the definition of the Christoffel symbols of the first kind is used in the second equality. On unifying k with i we obtain:
    $$\begin{aligned} \Gamma^i_{ji} &= \frac{1}{2g_{ii}} \left(\partial_i g_{ij} + \partial_j g_{ii} - \partial_i g_{ji} \right) \\ &= \frac{1}{2g_{ii}} \left(\partial_i g_{ij} + \partial_j g_{ii} - \partial_i g_{ij} \right) \\ &= \frac{1}{2g_{ii}} \partial_j g_{ii} \\ &= \frac{1}{2} \partial_j \ln g_{ii} \end{aligned}$$
    (with no sum over i) which is the required result. We note that line 2 is based on the symmetry of the metric tensor and line 4 is based on the rule of differentiation of natural logarithm.

16. Considering the identicality and difference of the indices of the Christoffel symbols of either kind, how many cases we have? List these cases.
    Answer: We have 4 main cases:
    (a) All the indices are identical.
    (b) Only two non-paired indices are identical.[19]

---

[19] In fact, we have two possibilities for this case but because of the symmetry of the Christoffel symbols in their paired indices they are combined in a single case.

5 TENSOR DIFFERENTIATION

(c) Only the two paired indices are identical.

(d) All the indices are different.

17. Prove the following relation which is used in Exercise 13: $\partial_i g = g g^{jl} \partial_i g_{jl}$.

    Answer:[20] According to the standard definition of determinant (which is given in any linear algebra text), the determinant g of the covariant metric tensor g_{ij} is given by:

    $$g = g_{ia} G^{ia}$$

    where G^{ia} is the cofactor of the entry g_{ia} and summation convention applies to a only since i represents a given row (or column). On taking the partial derivative of the two sides of the last equation with respect to the entry g_{ij} and using the product rule of differentiation we obtain:

    $$\frac{\partial g}{\partial g_{ij}} = G^{ia} \frac{\partial g_{ia}}{\partial g_{ij}} + g_{ia} \frac{\partial G^{ia}}{\partial g_{ij}}$$

    Now, according to the definition of the cofactor G^{ia} of the entry g_{ia}, G^{ia} contains no g_{ij} (i.e. G^{ia} is independent of g_{ij}) and hence the second term is zero because the partial derivative is zero, that is:

    $$\frac{\partial g}{\partial g_{ij}} = G^{ia} \frac{\partial g_{ia}}{\partial g_{ij}}$$

    Moreover, since the entries of a row (or a column) are independent of each other, we should have:[21]

    $$\frac{\partial g_{ia}}{\partial g_{ij}} = \delta_a^j$$

    Therefore, we have:

    $$\frac{\partial g}{\partial g_{ij}} = G^{ia} \delta_a^j = G^{ij}$$

    By the chain rule of differentiation we also have:

    $$\frac{\partial g}{\partial x^i} = \frac{\partial g}{\partial g_{ab}} \frac{\partial g_{ab}}{\partial x^i} = G^{ab} \frac{\partial g_{ab}}{\partial x^i}$$

    Now, the contravariant metric tensor is the inverse of the covariant metric tensor, and hence from the definition of matrix inverse we should have (noting the symmetry of the metric tensor):

    $$g^{ab} = \frac{G^{ab}}{g}$$

    On comparing the last two equations we conclude:

    $$\frac{\partial g}{\partial x^i} = g g^{ab} \frac{\partial g_{ab}}{\partial x^i}$$

---

[20] This answer is almost an exact replica (with some explanatory remarks) of the proof given in the Sokolnikoff book which I cited in the References of my book.

[21] An upper index in the denominator of partial derivative is like a lower index in the numerator, and hence a lower index in the denominator (i.e. j in g_{ij}) should be like an upper index in the numerator (i.e. j in δ_a^j). This similarly applies to the next equation.

5   TENSOR DIFFERENTIATION

On relabeling the dummy indices and using shorthand notation, we obtain:

$$\partial_i g = g g^{jl} \partial_i g_{jl}$$

which is the required result.

18. In orthogonal coordinate systems of a 3D space the number of independent non-identically vanishing Christoffel symbols of either kind is only 15. Explain why.
   Answer: In a 3D space we have 27 Christoffel symbols of either kind representing all the possible permutations of the three indices including the repetitive ones. In orthogonal coordinate systems the Christoffel symbols of either kind vanish when the indices are all different. This was established in Exercise 14 for the first kind, and when the first kind vanishes the second kind also vanishes according to the definition of the second kind which was given earlier. Hence, out of a total of 27 symbols, only 21 non-identically vanishing symbols are left since the 6 non-repetitive permutations are dropped. Now, since the Christoffel symbols are symmetric in their paired indices, then only 15 independent non-identically vanishing symbols will remain since 6 other permutations representing these symmetric exchanges are also dropped because they are not independent.

19. Verify the following equations related to the Christoffel symbols in orthogonal coordinate systems in a 3D space:

$$[12,1] = h_1 h_{1,2} \qquad \Gamma_{23}^3 = \frac{h_{3,2}}{h_3}$$

   Answer: The first equation is verified in the main text and hence, instead of repeating we verify another entry, say $[22,1] = -h_2 h_{2,1}$, that is:

$$\begin{aligned} [22,1] &= -\frac{1}{2}\partial_1 g_{22} \\ &= -\frac{1}{2}\partial_1 (h_2)^2 \\ &= -\frac{1}{2} 2 h_2 \partial_1 h_2 \\ &= -h_2 \partial_1 h_2 \\ &= -h_2 h_{2,1} \end{aligned}$$

   where line 1 is justified by the relation $[ii,j] = -\frac{1}{2}\partial_j g_{ii}$ in orthogonal systems ($i \neq j$, no sum on i) which is given in the main text, line 2 is justified by the relation $g_{ii} = (h_i)^2$ in orthogonal systems (no sum on i), line 3 is justified by the rules of differentiation, and lines 4 and 5 are simple algebraic manipulation and notation.
   Regarding the second equation, we have:

$$\begin{aligned} \Gamma_{23}^3 &= \frac{1}{2 g_{33}} \partial_2 g_{33} \\ &= \frac{1}{2(h_3)^2} \partial_2 (h_3)^2 \end{aligned}$$

$$= \frac{2h_3}{2(h_3)^2}\partial_2 h_3$$
$$= \frac{1}{h_3}\partial_2 h_3$$
$$= \frac{h_{3,2}}{h_3}$$

where line 1 is justified by the relation $\Gamma^i_{ji} = \frac{1}{2g_{ii}}\partial_j g_{ii}$ in orthogonal systems (no sum on i) which is given in the main text, line 2 is justified by the relation $g_{ii} = (h_i)^2$ in orthogonal systems (no sum on i), line 3 is justified by the rules of differentiation, and lines 4 and 5 are simple algebraic manipulation and notation.

20. Justify the following statement: "In any coordinate system, all the Christoffel symbols of either kind vanish identically *iff* all the components of the metric tensor in the given coordinate system are constants".

    Answer: From the definition of the Christoffel symbols, i.e.

$$[ij,k] = \frac{1}{2}(\partial_j g_{ik} + \partial_i g_{jk} - \partial_k g_{ij})$$
$$\Gamma^k_{ij} = \frac{g^{kl}}{2}(\partial_j g_{il} + \partial_i g_{jl} - \partial_l g_{ij})$$

we can see that both kinds are sum of terms containing partial derivatives of components of the metric tensor. Therefore, if all the components of the metric tensor are constants then all these partial derivatives will vanish identically and hence the Christoffel symbols will also vanish identically. Similarly, if we consider the general case of curvilinear systems then when the Christoffel symbols vanish identically then the individual partial derivatives must vanish identically and hence all the components of the metric tensor must be constants. In other words, if the Christoffel symbols of the first kind vanished identically but the individual partial derivatives did not vanish identically then we should have:

$$\partial_j g_{ik} + \partial_i g_{jk} = \partial_k g_{ij}$$

which cannot be true in general since the components of the metric tensor in the above equation are independent of each other. This argument similarly applies to the Christoffel symbols of the second kind, as can be seen from the above definition of Γ^k_{ij} (noting also that g^{kl} cannot vanish identically).[22]

21. Using the definition of the Christoffel symbols of the first kind with the metric tensor of the cylindrical coordinate system, find the Christoffel symbols of the first kind corresponding to the Euclidean metric of cylindrical coordinate systems.

---

[22] The more formal way of proving the "only if" part (i.e. if the Christoffel symbols vanish identically then all the components of the metric tensor in the given coordinate system are constants) is to use the fact that when the Christoffel symbols vanish identically the covariant derivative becomes partial derivative. Now, by the Ricci theorem, the covariant derivative of the metric tensor is zero and hence the partial derivative is zero in this case. Accordingly, the metric tensor must be constant, as required. However, we think our argument is simple, sufficient and more clear.

Answer: As explained earlier (see Exercise 18), we should have 27 symbols. However, because cylindrical coordinate systems are orthogonal then we should have only 15 independent non-identically vanishing symbols which are:[23]

$$[11,1] = \frac{1}{2}(\partial_1 g_{11} + \partial_1 g_{11} - \partial_1 g_{11}) = \frac{1}{2}(\partial_\rho 1 + \partial_\rho 1 - \partial_\rho 1) = 0$$

$$[11,2] = \frac{1}{2}(\partial_1 g_{12} + \partial_1 g_{12} - \partial_2 g_{11}) = \frac{1}{2}(\partial_\rho 0 + \partial_\rho 0 - \partial_\phi 1) = 0$$

$$[11,3] = \frac{1}{2}(\partial_1 g_{13} + \partial_1 g_{13} - \partial_3 g_{11}) = \frac{1}{2}(\partial_\rho 0 + \partial_\rho 0 - \partial_z 1) = 0$$

$$[12,1] = \frac{1}{2}(\partial_2 g_{11} + \partial_1 g_{21} - \partial_1 g_{12}) = \frac{1}{2}(\partial_\phi 1 + \partial_\rho 0 - \partial_\rho 0) = 0 = [21,1]$$

$$[12,2] = \frac{1}{2}(\partial_2 g_{12} + \partial_1 g_{22} - \partial_2 g_{12}) = \frac{1}{2}(\partial_\phi 0 + \partial_\rho \rho^2 - \partial_\phi 0) = \rho = [21,2]$$

$$[13,1] = \frac{1}{2}(\partial_3 g_{11} + \partial_1 g_{31} - \partial_1 g_{13}) = \frac{1}{2}(\partial_z 1 + \partial_\rho 0 - \partial_\rho 0) = 0 = [31,1]$$

$$[13,3] = \frac{1}{2}(\partial_3 g_{13} + \partial_1 g_{33} - \partial_3 g_{13}) = \frac{1}{2}(\partial_z 0 + \partial_\rho 1 - \partial_z 0) = 0 = [31,3]$$

$$[22,1] = \frac{1}{2}(\partial_2 g_{21} + \partial_2 g_{21} - \partial_1 g_{22}) = \frac{1}{2}(\partial_\phi 0 + \partial_\phi 0 - \partial_\rho \rho^2) = -\rho$$

$$[22,2] = \frac{1}{2}(\partial_2 g_{22} + \partial_2 g_{22} - \partial_2 g_{22}) = \frac{1}{2}(\partial_\phi \rho^2 + \partial_\phi \rho^2 - \partial_\phi \rho^2) = 0$$

$$[22,3] = \frac{1}{2}(\partial_2 g_{23} + \partial_2 g_{23} - \partial_3 g_{22}) = \frac{1}{2}(\partial_\phi 0 + \partial_\phi 0 - \partial_z \rho^2) = 0$$

$$[23,2] = \frac{1}{2}(\partial_3 g_{22} + \partial_2 g_{32} - \partial_2 g_{23}) = \frac{1}{2}(\partial_z \rho^2 + \partial_\phi 0 - \partial_\phi 0) = 0 = [32,2]$$

$$[23,3] = \frac{1}{2}(\partial_3 g_{23} + \partial_2 g_{33} - \partial_3 g_{23}) = \frac{1}{2}(\partial_z 0 + \partial_\phi 1 - \partial_z 0) = 0 = [32,3]$$

$$[33,1] = \frac{1}{2}(\partial_3 g_{31} + \partial_3 g_{31} - \partial_1 g_{33}) = \frac{1}{2}(\partial_z 0 + \partial_z 0 - \partial_\rho 1) = 0$$

$$[33,2] = \frac{1}{2}(\partial_3 g_{32} + \partial_3 g_{32} - \partial_2 g_{33}) = \frac{1}{2}(\partial_z 0 + \partial_z 0 - \partial_\phi 1) = 0$$

$$[33,3] = \frac{1}{2}(\partial_3 g_{33} + \partial_3 g_{33} - \partial_3 g_{33}) = \frac{1}{2}(\partial_z 1 + \partial_z 1 - \partial_z 1) = 0$$

while the remaining 6 symbols are identically zero.

22. Give all the Christoffel symbols of the first and second kind of the following coordinate systems: orthonormal Cartesian, cylindrical and spherical.
Answer:
Orthonormal Cartesian: all symbols of both kinds are zero.

---

[23] We note that ∂_1, ∂_2, and ∂_3 mean ∂_ρ, ∂_ϕ, and ∂_z. We also note that in cylindrical systems we have $g_{11} = 1$, $g_{22} = \rho^2$ and $g_{33} = 1$ while all the other entries are zero. Also, "non-identically vanishing" here means from the perspective of orthogonal systems although some are identically vanishing from the perspective of cylindrical systems.

Cylindrical: all symbols of the first kind are zero except the following:

$$[22, 1] = -\rho$$
$$[12, 2] = [21, 2] = \rho$$

All symbols of the second kind are zero except the following:

$$\Gamma^1_{22} = -\rho$$
$$\Gamma^2_{12} = \Gamma^2_{21} = \frac{1}{\rho}$$

Spherical: all symbols of the first kind are zero except the following:

$$[22, 1] = -r$$
$$[33, 1] = -r\sin^2\theta$$
$$[12, 2] = [21, 2] = r$$
$$[33, 2] = -r^2 \sin\theta\cos\theta$$
$$[13, 3] = [31, 3] = r\sin^2\theta$$
$$[23, 3] = [32, 3] = r^2 \sin\theta\cos\theta$$

All symbols of the second kind are zero except the following:

$$\Gamma^1_{22} = -r$$
$$\Gamma^1_{33} = -r\sin^2\theta$$
$$\Gamma^2_{12} = \Gamma^2_{21} = \frac{1}{r}$$
$$\Gamma^2_{33} = -\sin\theta\cos\theta$$
$$\Gamma^3_{13} = \Gamma^3_{31} = \frac{1}{r}$$
$$\Gamma^3_{23} = \Gamma^3_{32} = \cot\theta$$

23. Mention two important properties of the Christoffel symbols of either kind with regard to the order and similarity of their indices.
    Answer: One property is the symmetry of these symbols in their paired indices, i.e. $[ij, k] = [ji, k]$ and $\Gamma^k_{ij} = \Gamma^k_{ji}$. Another property is that in orthogonal systems these symbols vanish identically when all their indices are different, i.e. $[ij, k] = 0$ and $\Gamma^k_{ij} = 0$ when $i \neq j \neq k$.

24. Using the analytical expressions of the Christoffel symbols of the second kind in orthogonal systems plus the entries of the table of scale factors (which is given in the book) and the properties of the Christoffel symbols of the second kind, derive these symbols corresponding to the metrics of the coordinate systems of question 22.
    Answer:
    Orthonormal Cartesian: all the scale factors are 1. Moreover, all the analytical expressions of the Christoffel symbols of the second kind in orthogonal systems contain

derivatives of the scale factors and these derivatives must be zero. Hence, all these symbols are zero.

Cylindrical: the scale factors are $h_1 = h_3 = 1$ and $h_2 = \rho$. Hence, all these symbols must be zero except those whose analytical expression contains derivative of h_2 with respect to ρ, i.e. $h_{2,1}$, that is:

$$\Gamma^2_{12} = +\frac{h_{2,1}}{h_2} = \frac{\partial_\rho \rho}{\rho} = \frac{1}{\rho} = \Gamma^2_{21}$$

$$\Gamma^1_{22} = -\frac{h_2 h_{2,1}}{(h_1)^2} = -\frac{\rho \partial_\rho \rho}{1^2} = -\rho$$

Spherical: the scale factors are $h_1 = 1$, $h_2 = r$ and $h_3 = r\sin\theta$. Hence, all these symbols must be zero except those whose analytical expression contains derivative of h_2 with respect to r (i.e. $h_{2,1}$) or derivative of h_3 with respect to r (i.e. $h_{3,1}$) or derivative of h_3 with respect to θ (i.e. $h_{3,2}$), that is:

$$\Gamma^2_{12} = +\frac{h_{2,1}}{h_2} = \frac{\partial_r r}{r} = \frac{1}{r} = \Gamma^2_{21}$$

$$\Gamma^1_{22} = -\frac{h_2 h_{2,1}}{(h_1)^2} = -\frac{r \partial_r r}{1^2} = -r$$

$$\Gamma^3_{13} = +\frac{h_{3,1}}{h_3} = \frac{\partial_r (r\sin\theta)}{r\sin\theta} = \frac{\sin\theta}{r\sin\theta} = \frac{1}{r} = \Gamma^3_{31}$$

$$\Gamma^1_{33} = -\frac{h_3 h_{3,1}}{(h_1)^2} = -\frac{r\sin\theta\, \partial_r (r\sin\theta)}{1^2} = -r\sin^2\theta$$

$$\Gamma^3_{23} = +\frac{h_{3,2}}{h_3} = \frac{\partial_\theta (r\sin\theta)}{r\sin\theta} = \frac{r\cos\theta}{r\sin\theta} = \cot\theta = \Gamma^3_{32}$$

$$\Gamma^2_{33} = -\frac{h_3 h_{3,2}}{(h_2)^2} = -\frac{r\sin\theta\, \partial_\theta (r\sin\theta)}{r^2} = -\frac{r\sin\theta\, r\cos\theta}{r^2} = -\sin\theta\cos\theta$$

25. Write the following Christoffel symbols in terms of the coordinates instead of the indices assuming a cylindrical system: $[12, 1]$, $[23, 1]$, Γ^2_{21} and Γ^3_{32}. Do the same assuming a spherical system.
    Answer:
    Cylindrical: $[\rho\phi, \rho]$, $[\phi z, \rho]$, $\Gamma^\phi_{\phi\rho}$ and $\Gamma^z_{z\phi}$.
    Spherical: $[r\theta, r]$, $[\theta\phi, r]$, $\Gamma^\theta_{\theta r}$ and $\Gamma^\phi_{\phi\theta}$.

26. Show that all the Christoffel symbols will vanish when the components of the metric tensor are constants.
    Answer: The Christoffel symbols of the first and second kind are defined as.

$$[ij, k] = \frac{1}{2} (\partial_j g_{ik} + \partial_i g_{jk} - \partial_k g_{ij})$$

$$\Gamma^k_{ij} = \frac{g^{kl}}{2} (\partial_j g_{il} + \partial_i g_{jl} - \partial_l g_{ij})$$

As we see, both kinds are sum of terms containing partial derivatives of components of the metric tensor. Therefore, when all the components of the metric tensor are constants all these partial derivatives will vanish identically and hence the Christoffel symbols will also vanish identically.

27. Why the Christoffel symbols of either kind may be superscripted or subscripted by the symbol of the underlying metric tensor? When this (or other measures for indicating the underlying metric tensor) becomes necessary? Mention some of the other measures used to indicate the underlying metric tensor.
Answer: The purpose of the superscripts and subscripts is to indicate the metric from which the Christoffel symbols are derived (see the definition of the Christoffel symbols which is given in the answer of the previous question).
This becomes necessary when in a given context we have two or more different metrics related to two or more different spaces or systems and hence the metric of the symbols requires clarification.
Indicating the underlying metric tensor of the Christoffel symbols may also be done by using different types of indices for each metric, e.g. by using Latin or upper case indices for the Christoffel symbols of one metric and Greek or lower case indices for the Christoffel symbols of the other metric.

28. Explain why the total number of independent Christoffel symbols of each kind is equal to $\frac{n^2(n+1)}{2}$ where n is the dimension of the space.[24]
Answer: Considering the identicality and difference of the indices of the Christoffel symbols of either kind in general coordinate systems, we have 4 main cases:
(a) All the indices are identical: this represents n independent symbols since we have n values for any index.
(b) Only two non-paired indices are identical: this represents $n(n-1)$ independent symbols, i.e. n identical times $(n-1)$ different (or the other way around).[25]
(c) Only the two paired indices are identical: this represents $n(n-1)$ independent symbols, i.e. n non-paired times $(n-1)$ paired (or the other way around).
(d) All the indices are different: this represents $n(n-1)(n-2)$ symbols which is the number of non-repetitive permutations. However, due to the symmetry in the paired indices we have only $\frac{n(n-1)(n-2)}{2}$ independent symbols.
Accordingly, the total number of independent symbols is:

$$\begin{aligned} N_{\text{CI}} &= n + n(n-1) + n(n-1) + \frac{n(n-1)(n-2)}{2} \\ &= \frac{2n + 2n(n-1) + 2n(n-1) + n(n-1)(n-2)}{2} \\ &= \frac{2n + 2n^2 - 2n + 2n^2 - 2n + n^3 - 2n^2 - n^2 + 2n}{2} \end{aligned}$$

---

[24] The reader should note that this question is about general coordinate systems and hence it should not be confused with certain special coordinate systems like orthogonal systems.

[25] As indicated before (see the footnote of Exercise 16), we have two possibilities for this case where for each possibility we have $\frac{n(n-1)}{2}$ independent symbols and hence the total number of independent symbols in this case is $n(n-1)$.

5   TENSOR DIFFERENTIATION   97

$$= \frac{n^3 + n^2}{2}$$

$$= \frac{n^2(n+1)}{2}$$

29. Why covariant differentiation of tensors is regarded as a generalization of the ordinary partial differentiation?
    Answer: Because the ordinary partial differentiation of tensors (i.e. applied to their components only) is valid only in certain types of coordinate systems (affine or rectilinear) where the basis vectors are coordinate-independent and hence in general curvilinear coordinate systems where the basis vectors are coordinate-dependent, the ordinary partial differentiation of tensors does not produce tensors in general (i.e. it does not satisfy the invariance principle of tensors). In contrast, the covariant differentiation of tensors necessarily produces tensors and hence it is a generalization of the ordinary partial differentiation since it is valid in general while ordinary partial differentiation is valid only in particular coordinate systems. This also applies to absolute differentiation as a generalization of the ordinary total differentiation.

30. In general curvilinear coordinate systems, the variation of the basis vectors should also be considered in the differentiation process of non-scalar tensors. Why?
    Answer: Because in general curvilinear coordinate systems the basis vectors, as well as the components, depend on the coordinates in general and hence they are variable functions of coordinates. Therefore, by the product rule of differentiation, applied to non-scalar tensors in curvilinear systems, both the components and the basis vectors should be subjected to the differentiation process to take account of the variation of both parts (i.e. components and basis vectors) of the tensor because a non-scalar tensor is made of components multiplied by basis vectors. In brief, ordinary differentiation is justified in rectilinear systems because the basis vectors are constants and hence all we need to do (according to the product rule of differentiation) is to differentiate the components, but in curvilinear coordinate systems this process (i.e. differentiation of components only) is only part of the differentiation process and hence it is not sufficient for differentiating general tensors. Therefore, the other part of the differentiation process according to the product rule should be added and this complete differentiation process (in which both the components and the basis vectors are differentiated) is what is called tensor differentiation (i.e. covariant and absolute differentiation).

31. State the mathematical definition of contravariant differentiation of a tensor A_i.
    Answer: Contravariant differentiation is achieved by raising the differentiation index of the covariant derivative using the index raising operator. Hence, the contravariant differentiation of a covariant vector A_i, for example, is given by:

    $$A_i^{;j} = g^{jk} A_{i;k}$$

    where g^{jk} is the contravariant metric tensor.

32. Obtain analytical expressions for $A_{i;j}$ and $B^i_{;j}$ by differentiating the vectors $\mathbf{A} = A_i \mathbf{E}^i$ and $\mathbf{B} = B^i \mathbf{E}_i$.

Answer: We have:

$$\begin{aligned}
\partial_j \left(A_i \mathbf{E}^i\right) &= \mathbf{E}^i \partial_j A_i + A_i \partial_j \mathbf{E}^i \\
&= \mathbf{E}^i \partial_j A_i - A_i \Gamma^i_{kj} \mathbf{E}^k \\
&= \mathbf{E}^i \partial_j A_i - A_k \Gamma^k_{ij} \mathbf{E}^i \\
&= \left(\partial_j A_i - A_k \Gamma^k_{ij}\right) \mathbf{E}^i \\
&= A_{i;j} \mathbf{E}^i
\end{aligned}$$

where in line 1 we use the product rule of differentiation since both the components and basis vectors of general tensors are coordinate-dependent variables, in line 2 we use the identity $\partial_j \mathbf{E}^i = -\Gamma^i_{kj} \mathbf{E}^k$ which is given in the book, in line 3 we relabel the indices i and k, in line 4 we take out the common factor $\mathbf{E}^i$, and in line 5 we use a shorthand notation for the covariant derivative which is based on its definition.

Similarly:

$$\begin{aligned}
\partial_j \left(B^i \mathbf{E}_i\right) &= \mathbf{E}_i \partial_j B^i + B^i \partial_j \mathbf{E}_i \\
&= \mathbf{E}_i \partial_j B^i + B^i \Gamma^k_{ij} \mathbf{E}_k \\
&= \mathbf{E}_i \partial_j B^i + B^k \Gamma^i_{kj} \mathbf{E}_i \\
&= \left(\partial_j B^i + B^k \Gamma^i_{kj}\right) \mathbf{E}_i \\
&= B^i_{;j} \mathbf{E}_i
\end{aligned}$$

where in line 2 we use the identity $\partial_j \mathbf{E}_i = \Gamma^k_{ij} \mathbf{E}_k$ which is given in the book, while the other lines are similarly justified as in the first part.

33. Repeat question 32 with the rank-2 tensors $\mathbf{C} = C_{ij} \mathbf{E}^i \mathbf{E}^j$ and $\mathbf{D} = D^{ij} \mathbf{E}_i \mathbf{E}_j$ to obtain $C_{ij;k}$ and $D^{ij}_{;k}$.

Answer: We have:

$$\begin{aligned}
\partial_k \left(C_{ij} \mathbf{E}^i \mathbf{E}^j\right) &= \mathbf{E}^i \mathbf{E}^j \partial_k C_{ij} + C_{ij} \left(\partial_k \mathbf{E}^i\right) \mathbf{E}^j + C_{ij} \mathbf{E}^i \left(\partial_k \mathbf{E}^j\right) \\
&= \mathbf{E}^i \mathbf{E}^j \partial_k C_{ij} - C_{ij} \left(\Gamma^i_{ak} \mathbf{E}^a\right) \mathbf{E}^j - C_{ij} \mathbf{E}^i \left(\Gamma^j_{ak} \mathbf{E}^a\right) \\
&= \mathbf{E}^i \mathbf{E}^j \partial_k C_{ij} - C_{aj} \Gamma^a_{ik} \mathbf{E}^i \mathbf{E}^j - C_{ia} \Gamma^a_{jk} \mathbf{E}^i \mathbf{E}^j \\
&= \left(\partial_k C_{ij} - C_{aj} \Gamma^a_{ik} - C_{ia} \Gamma^a_{jk}\right) \mathbf{E}^i \mathbf{E}^j \\
&= C_{ij;k} \mathbf{E}^i \mathbf{E}^j
\end{aligned}$$

where the lines are justified as in the answer of the previous question.

Similarly, we have:

$$\begin{aligned}
\partial_k \left(D^{ij} \mathbf{E}_i \mathbf{E}_j\right) &= \mathbf{E}_i \mathbf{E}_j \partial_k D^{ij} + D^{ij} \left(\partial_k \mathbf{E}_i\right) \mathbf{E}_j + D^{ij} \mathbf{E}_i \left(\partial_k \mathbf{E}_j\right) \\
&= \mathbf{E}_i \mathbf{E}_j \partial_k D^{ij} + D^{ij} \left(\Gamma^a_{ik} \mathbf{E}_a\right) \mathbf{E}_j + D^{ij} \mathbf{E}_i \left(\Gamma^a_{jk} \mathbf{E}_a\right) \\
&= \mathbf{E}_i \mathbf{E}_j \partial_k D^{ij} + D^{aj} \Gamma^i_{ak} \mathbf{E}_i \mathbf{E}_j + D^{ia} \Gamma^j_{ak} \mathbf{E}_i \mathbf{E}_j \\
&= \left(\partial_k D^{ij} + D^{aj} \Gamma^i_{ak} + D^{ia} \Gamma^j_{ak}\right) \mathbf{E}_i \mathbf{E}_j \\
&= D^{ij}_{;k}
\end{aligned}$$

5 *TENSOR DIFFERENTIATION*   99

34. For a differentiable tensor **A** of type (m, n), the covariant derivative with respect to the coordinate u^k is given by:

$$A^{i_1 i_2 \ldots i_m}_{j_1 j_2 \ldots j_n; k} = \partial_k A^{i_1 i_2 \ldots i_m}_{j_1 j_2 \ldots j_n} + \Gamma^{i_1}_{lk} A^{l i_2 \ldots i_m}_{j_1 j_2 \ldots j_n} + \Gamma^{i_2}_{lk} A^{i_1 l \ldots i_m}_{j_1 j_2 \ldots j_n} + \cdots + \Gamma^{i_m}_{lk} A^{i_1 i_2 \ldots l}_{j_1 j_2 \ldots j_n}$$
$$- \Gamma^{l}_{j_1 k} A^{i_1 i_2 \ldots i_m}_{l j_2 \ldots j_n} - \Gamma^{l}_{j_2 k} A^{i_1 i_2 \ldots i_m}_{j_1 l \ldots j_n} - \cdots - \Gamma^{l}_{j_n k} A^{i_1 i_2 \ldots i_m}_{j_1 j_2 \ldots l}$$

Extract from the pattern of this expression the practical rules that should be followed in writing the analytical expressions of covariant derivative of tensors of any rank and type.
Answer: The practical rules that can be extracted from the pattern of this expression are summarized in the following where we call the indices $i_1 i_2 \ldots i_m$ and $j_1 j_2 \ldots j_n$ the basis indices (since each one of these indices corresponds to one basis vector of the basis tensor[26]) and call k the differentiation index.
We start with an ordinary partial derivative term of the component of the given tensor with respect to the differentiation index (i.e. the first term on the right hand side of the above equation). Then for each basis index of the tensor we add an extra Christoffel symbol term where this term satisfies the following properties:
• The term is positive if the basis index is contravariant and negative if the basis index is covariant.
• One of the lower indices of the Christoffel symbol is the differentiation index.
• The basis index of the tensor in the concerned term is contracted with one of the indices of the Christoffel symbol using a new label (i.e. the new label is not in use already as a basis index) and hence they are opposite in their variance type.
• The label of the basis index of that term is transferred from the tensor to the Christoffel symbol keeping its position as lower or upper.
• All the other indices of the tensor in the concerned term keep their labels, position and order.

35. In the expression of covariant derivative, what the partial derivative term stands for and what the Christoffel symbol terms represent?
Answer: The partial derivative term stands for the derivative of the component of the tensor, while each Christoffel symbol term stands for the derivative of one basis vector of the basis tensor of the tensor (e.g. the basis tensor of A_{ij} is $\mathbf{E}^i \mathbf{E}^j$). More clearly, the partial derivative term represents the rate of change of the tensor component with change of position as a result of moving along the coordinate curve of the differentiation index, while the Christoffel symbol terms represent the change experienced by the local basis vectors as a result of the same movement.

36. For the covariant derivative of a type (m, n, w) tensor, obtain the total number of terms, the number of negative Christoffel symbol terms and the number of positive Christoffel symbol terms.
Answer: We have 1 partial derivative term and $m + n$ Christoffel symbol terms and hence the total number of terms is $m + n + 1$. We have n covariant indices and hence

---
[26] For example, the basis tensor of A_{ij} is $\mathbf{E}^i \mathbf{E}^j$.

the number of negative Christoffel symbol terms is n. We have m contravariant indices and hence the number of positive Christoffel symbol terms is m.

37. What is the rank and type of the covariant derivative of a tensor of rank-n and type (p,q)?
    Answer: The covariant derivative of a tensor is a tensor whose covariant rank is higher than the covariant rank of the original tensor by one. Hence, the rank is $p+q+1$ (or $n+1$) and the type is $(p, q+1)$.

38. The covariant derivative of a differentiable scalar function is the same as the ordinary partial derivative. Why?
    Answer: Because the scalar is not referred to any basis vector or basis tensor and hence it has no Christoffel symbol terms, so what remains of the covariant derivative terms is the partial derivative term only and hence the covariant derivative of a scalar is the same as the ordinary partial derivative.

39. What is the significance of the dependence of the covariant derivative on the Christoffel symbols with regard to its relation to the space and coordinate system?
    Answer: As seen earlier, the Christoffel symbols are solely dependent on the metric tensor and hence they can be seen as a characteristic property of both the underlying space and the employed coordinate system. So, the dependency of the covariant derivative on the Christoffel symbols implies its dependency on the space and the coordinate system and hence it is characterized by the space and the coordinate system and it reflects their features.

40. The covariant derivative of tensors in coordinate systems with constant basis vectors is the same as the ordinary partial derivative for all tensor ranks. Why?
    Answer: Because the Christoffel symbol terms, which represent the derivatives of the basis vectors, will vanish since the basis vectors are constant. So, what remains of the covariant derivative terms is the partial derivative term only and hence the covariant derivative of tensors in coordinate systems with constant basis vectors is the same as the ordinary partial derivative for all tensor ranks.

41. Express, mathematically, the fact that the metric tensor is in lieu of constant with respect to covariant differentiation.
    Answer: This can be expressed in several forms such as:
    $$\begin{aligned} \partial_{;k} g_{ij} &= 0 \\ \partial_{;k} g^{ij} &= 0 \\ \partial_{;k} \left(g_{ij} A^j \right) &= g_{ij} \partial_{;k} A^j \\ \mathbf{g}_{;k} &= \mathbf{0} \\ (\mathbf{g} \circ \mathbf{A})_{;k} &= \mathbf{g} \circ \mathbf{A}_{;k} \end{aligned}$$
    where $\mathbf{g}$ is the metric tensor in symbolic notation and the symbol $\circ$ denotes an inner or outer product operator.

42. Which rules of ordinary partial differentiation also apply to covariant differentiation and which rules do not? State all these rules symbolically for both ordinary and covariant differentiation.

5 TENSOR DIFFERENTIATION

Answer:
Linearity: this applies to both ordinary partial differentiation and covariant differentiation, that is:

$$\frac{\partial}{\partial x}(af+bg) = a\frac{\partial f}{\partial x} + b\frac{\partial g}{\partial x}$$
$$(a\mathbf{A} \pm b\mathbf{B})_{;i} = a\mathbf{A}_{;i} \pm b\mathbf{B}_{;i}$$

where a and b are scalar constants, f and g are differentiable scalar functions and $\mathbf{A}$ and $\mathbf{B}$ are differentiable tensors.

The product rule of differentiation: this applies to both ordinary partial differentiation and covariant differentiation, that is:

$$\frac{\partial}{\partial x}(fg) = g\frac{\partial f}{\partial x} + f\frac{\partial g}{\partial x}$$
$$(\mathbf{A} \circ \mathbf{B})_{;i} = \mathbf{A}_{;i} \circ \mathbf{B} + \mathbf{A} \circ \mathbf{B}_{;i}$$

where the symbol $\circ$ denotes an inner or outer product operator. However, the order of the tensors in the covariant differentiation should be observed.

Commutativity of operators: this applies to ordinary partial differentiation but not to covariant differentiation, that is:

$$\partial_i \partial_j = \partial_j \partial_i$$
$$\partial_{;i} \partial_{;j} \neq \partial_{;j} \partial_{;i}$$

43. Explain why the covariant differential operators with respect to different indices do not commute, i.e. $\partial_{;i}\partial_{;j} \neq \partial_{;j}\partial_{;i}$ ($i \neq j$).

    Answer: The reason is that the differentiation indices, like the indices of the differentiated tensor, are referred to basis vectors and since the basis vectors do not commute then the differentiation indices (and hence the covariant differential operators that represent these indices) do not commute. This is unlike the indices of the ordinary partial differential operators since these indices are not referred to basis vectors. In brief, the covariant derivative of a tensor is a tensor whose covariant rank increases by 1 for each covariant differentiation operation and hence the differentiation indices are not different from the indices of the differentiated tensor. Therefore, the differentiation indices should follow the same rules that the indices of the differentiated tensor follow and one of these rules is the importance of the order of indices (i.e. non-commutativity of tensor indices). The non-commutativity of covariant differential operators can also be shown directly by performing the operations of covariant differentiation in different orders where it can be easily verified that the result depends on the order of the operators (see Exercises 52 and 53).

44. State the Ricci theorem about covariant differentiation of the metric tensor and prove it with full justification of each step.

    Answer: The Ricci theorem states that the covariant derivative of the metric tensor

is zero.
Covariant metric tensor:

$$
\begin{aligned}
\partial_{;k} g_{ij} &= \partial_{;k} \left(\mathbf{E}_i \cdot \mathbf{E}_j \right) \\
&= \left(\partial_{;k} \mathbf{E}_i \right) \cdot \mathbf{E}_j + \mathbf{E}_i \cdot \left(\partial_{;k} \mathbf{E}_j \right) \\
&= \mathbf{0} \cdot \mathbf{E}_j + \mathbf{E}_i \cdot \mathbf{0} \\
&= 0
\end{aligned}
$$

where line 1 is the relation between the metric tensor and the basis vectors, line 2 is the product rule of differentiation which applies to covariant differentiation as to ordinary differentiation, and line 3 is the fact that the covariant derivative of the basis vectors is identically zero (as demonstrated in the book and in Exercise 57).

Contravariant metric tensor:

$$
\begin{aligned}
\partial_{;k} g^{ij} &= \partial_{;k} \left(\mathbf{E}^i \cdot \mathbf{E}^j \right) \\
&= \left(\partial_{;k} \mathbf{E}^i \right) \cdot \mathbf{E}^j + \mathbf{E}^i \cdot \left(\partial_{;k} \mathbf{E}^j \right) \\
&= \mathbf{0} \cdot \mathbf{E}^j + \mathbf{E}^i \cdot \mathbf{0} \\
&= 0
\end{aligned}
$$

where the lines are similarly justified as in the covariant case.

Mixed metric tensor: the proof can be similar to the proof of the covariant and contravariant types, i.e. $\partial_{;k} g_i^j = \partial_{;k} \left(\mathbf{E}_i \cdot \mathbf{E}^j \right) = \cdots$ etc. Alternatively:

$$
\begin{aligned}
\partial_{;k} g_i^j &= \partial_{;k} \left(g_{ia} g^{aj} \right) \\
&= \left(\partial_{;k} g_{ia} \right) g^{aj} + g_{ia} \left(\partial_{;k} g^{aj} \right) \\
&= 0 + 0 \\
&= 0
\end{aligned}
$$

where line 1 is based on the use of index shifting operator, line 2 is the product rule of differentiation, and line 3 is based on the results that we already obtained in this question for the covariant and contravariant types.

45. State, symbolically, the commutative property of the covariant derivative operator with the index shifting operator (which is based on the Ricci theorem) using the symbolic notation one time and the indicial notation another.
Answer:
Symbolic notation:
$$\partial_{;k} \left(\mathbf{g} \cdot \mathbf{A} \right) = \mathbf{g} \cdot \left(\partial_{;k} \mathbf{A} \right)$$

Indicial notation:
$$\partial_{;k} \left(g_{ij} A^j \right) = g_{ij} \left(\partial_{;k} A^j \right) \qquad \text{or} \qquad \partial_{;k} \left(g^{ij} A_j \right) = g^{ij} \left(\partial_{;k} A_j \right)$$

46. Verify that the ordinary Kronecker delta tensor is constant with respect to covariant differentiation.

5 TENSOR DIFFERENTIATION

Answer:
Mixed type: we have:

$$\begin{aligned}\partial_{;k}\delta^i_j &= \partial_k \delta^i_j + \delta^a_j \Gamma^i_{ak} - \delta^i_a \Gamma^a_{jk} \\ &= 0 + \delta^a_j \Gamma^i_{ak} - \delta^i_a \Gamma^a_{jk} \\ &= 0 + \Gamma^i_{jk} - \Gamma^i_{jk} \\ &= 0 \end{aligned}$$

where line 1 is based on the rules of covariant derivative, line 2 is justified by the fact that all the components of the Kronecker delta are constant (i.e. either 0 or 1), and line 3 is based on using the Kronecker delta as an index replacement operator.
Covariant type: we have:

$$\begin{aligned}\partial_{;k}\delta_{ij} &= \partial_{;k}\left(g_{ia}\delta^a_j\right) \\ &= \left(\partial_{;k}g_{ia}\right)\delta^a_j + g_{ia}\left(\partial_{;k}\delta^a_j\right) \\ &= 0 + 0 \\ &= 0 \end{aligned}$$

where line 1 is based on using index shifting operator, line 2 is based on the product rule of differentiation which applies to covariant differentiation as to ordinary differentiation, and line 3 is based on the fact that the covariant derivative of the metric tensor and the covariant derivative of the mixed Kronecker delta are both zero (see Exercise 44 and the first part of the current question).
Contravariant type: we have:

$$\begin{aligned}\partial_{;k}\delta^{ij} &= \partial_{;k}\left(g^{ja}\delta^i_a\right) \\ &= \left(\partial_{;k}g^{ja}\right)\delta^i_a + g^{ja}\left(\partial_{;k}\delta^i_a\right) \\ &= 0 + 0 \\ &= 0 \end{aligned}$$

where the lines are justified as for the covariant type.

47. State, symbolically, the fact that covariant differentiation and contraction of index operations commute with each other.
Answer: We have:

$$\left(\partial_{;m} A^{ij}_k\right)\delta^k_j = \partial_{;m}\left(A^{ij}_k \delta^k_j\right)$$

where the left hand side is the order "covariant differentiation first followed by contraction of index" while the right hand side is the order "contraction of index first followed by covariant differentiation".

48. What is the condition on the components of the metric tensor that makes the covariant derivative become ordinary partial derivative for all tensor ranks?
Answer: The condition is that the components of the metric tensor are constants, because in this case all the Christoffel symbols will vanish identically (according to

their definition which was given earlier) and hence all the Christoffel symbol terms of the covariant derivative will vanish as well, so what remains of the covariant derivative terms is the partial derivative term only and hence the covariant derivative becomes an ordinary partial derivative for all tensor ranks.

49. Prove that covariant differentiation and contraction of indices commute.
   Answer: We have two cases:
   (a) Covariant differentiation first followed by contraction of index (i.e. contracting j with k in the following equation), that is:
   $$\left(\partial_{;m} A^{ij}_k\right) \delta^k_j = \left(A^{ij}_{k;m}\right) \delta^k_j = A^{ij}_{k;m} \delta^k_j = A^{ik}_{k;m}$$
   (b) Contraction of index first followed by covariant differentiation:
   $$\partial_{;m} \left(A^{ij}_k \delta^k_j\right) = \partial_{;m} \left(A^{ik}_k\right) = A^{ik}_{k;m}$$
   On comparing the two equations we obtain:
   $$\left(\partial_{;m} A^{ij}_k\right) \delta^k_j = \partial_{;m} \left(A^{ij}_k \delta^k_j\right)$$
   i.e. covariant differentiation and contraction of indices commute, as required.

50. What is the mathematical condition that is required if the mixed second order partial derivatives should be equal, i.e. $\partial_i \partial_j = \partial_j \partial_i$ $(i \neq j)$?
    Answer: It is the C^2 continuity condition, i.e. the differentiated function and all its first and second partial derivatives do exist and they are continuous in their domain.

51. What is the mathematical condition that is required if the mixed second order covariant derivatives should be equal, i.e. $\partial_{;i} \partial_{;j} = \partial_{;j} \partial_{;i}$ $(i \neq j)$?
    Answer: The condition is the vanishing of the Riemann-Christoffel curvature tensor.

52. Derive analytical expressions for $A_{i;jk}$ and $A_{i;kj}$ and hence verify that $A_{i;jk} \neq A_{i;kj}$.
    Answer: We have:
$$\begin{aligned}
A_{i;jk} &= (A_{i;j})_{;k} \\
&= \partial_k A_{i;j} - \Gamma^a_{ik} A_{a;j} - \Gamma^a_{jk} A_{i;a} \\
&= \partial_k \left(\partial_j A_i - \Gamma^b_{ij} A_b\right) - \Gamma^a_{ik} \left(\partial_j A_a - \Gamma^b_{aj} A_b\right) - \Gamma^a_{jk} \left(\partial_a A_i - \Gamma^b_{ia} A_b\right) \\
&= \partial_k \partial_j A_i - \Gamma^b_{ij} \partial_k A_b - A_b \partial_k \Gamma^b_{ij} - \Gamma^a_{ik} \partial_j A_a + \Gamma^a_{ik} \Gamma^b_{aj} A_b - \Gamma^a_{jk} \partial_a A_i + \Gamma^a_{jk} \Gamma^b_{ia} A_b
\end{aligned}$$
   where all these lines are justified by the definition of covariant differentiation and some other basic operations.
   Similarly, we have:
$$\begin{aligned}
A_{i;kj} &= (A_{i;k})_{;j} \\
&= \partial_j A_{i;k} - \Gamma^a_{ij} A_{a;k} - \Gamma^a_{kj} A_{i;a} \\
&= \partial_j \left(\partial_k A_i - \Gamma^b_{ik} A_b\right) - \Gamma^a_{ij} \left(\partial_k A_a - \Gamma^b_{ak} A_b\right) - \Gamma^a_{kj} \left(\partial_a A_i - \Gamma^b_{ia} A_b\right) \\
&= \partial_j \partial_k A_i - \Gamma^b_{ik} \partial_j A_b - A_b \partial_j \Gamma^b_{ik} - \Gamma^a_{ij} \partial_k A_a + \Gamma^a_{ij} \Gamma^b_{ak} A_b - \Gamma^a_{kj} \partial_a A_i + \Gamma^a_{kj} \Gamma^b_{ia} A_b
\end{aligned}$$

So, let take the difference between these two equations to see if $A_{i;jk} - A_{i;kj} = 0$ (and hence $A_{i;jk} = A_{i;kj}$) or $A_{i;jk} - A_{i;kj} \neq 0$ (and hence $A_{i;jk} \neq A_{i;kj}$), that is:

$$\begin{aligned}
& A_{i;jk} - A_{i;kj} \\
&= \partial_k \partial_j A_i - \Gamma^b_{ij} \partial_k A_b - A_b \partial_k \Gamma^b_{ij} - \Gamma^a_{ik} \partial_j A_a + \Gamma^a_{ik} \Gamma^b_{aj} A_b - \Gamma^a_{jk} \partial_a A_i + \Gamma^a_{jk} \Gamma^b_{ia} A_b - \\
&\quad \left(\partial_j \partial_k A_i - \Gamma^b_{ik} \partial_j A_b - A_b \partial_j \Gamma^b_{ik} - \Gamma^a_{ij} \partial_k A_a + \Gamma^a_{ij} \Gamma^b_{ak} A_b - \Gamma^a_{kj} \partial_a A_i + \Gamma^a_{kj} \Gamma^b_{ia} A_b \right) \\
&= -A_b \partial_k \Gamma^b_{ij} + \Gamma^a_{ik} \Gamma^b_{aj} A_b - \left(-A_b \partial_j \Gamma^b_{ik} + \Gamma^a_{ij} \Gamma^b_{ak} A_b \right) \\
&= -A_b \partial_k \Gamma^b_{ij} + \Gamma^a_{ik} \Gamma^b_{aj} A_b + A_b \partial_j \Gamma^b_{ik} - \Gamma^a_{ij} \Gamma^b_{ak} A_b \\
&= A_b R^b_{ijk}
\end{aligned}$$

where R^b_{ijk} is the Riemann-Christoffel curvature tensor.[27] Now, since this tensor does not vanish identically in general, then $A_{i;jk} - A_{i;kj} \neq 0$ in general and hence $A_{i;jk} \neq A_{i;kj}$, as required.

53. From the result of exercise 52 plus the definition of the Riemann-Christoffel curvature tensor of the second kind, verify the following relation: $A_{i;jk} - A_{i;kj} = A_b R^b_{ijk}$.
    Answer: The Riemann-Christoffel curvature tensor of the second kind is given by:

    $$R^i_{jkl} = \partial_k \Gamma^i_{jl} - \partial_l \Gamma^i_{jk} + \Gamma^r_{jl} \Gamma^i_{rk} - \Gamma^r_{jk} \Gamma^i_{rl}$$

    On inner multiplying with A_b and relabeling the indices we obtain:

    $$A_b R^b_{ijk} = A_b \partial_j \Gamma^b_{ik} - A_b \partial_k \Gamma^b_{ij} + A_b \Gamma^a_{ik} \Gamma^b_{aj} - A_b \Gamma^a_{ij} \Gamma^b_{ak}$$

    which is the same as the result of $A_{i;jk} - A_{i;kj}$ that we obtained in the previous exercise (i.e. the equation before the last in the answer of the previous question). Hence, we conclude that $A_{i;jk} - A_{i;kj} = A_b R^b_{ijk}$, as required.

54. What is the covariant derivative of a relative scalar f of weight w? What is the covariant derivative of a rank-2 relative tensor A^i_j of weight w?
    Answer: They are:

    $$\begin{aligned}
    \partial_{;k} f &= \partial_k f - w f \Gamma^a_{ak} \\
    \partial_{;k} A^i_j &= \partial_k A^i_j + \Gamma^i_{ak} A^a_j - \Gamma^a_{jk} A^i_a - w A^i_j \Gamma^a_{ak}
    \end{aligned}$$

55. Why the covariant derivative of a non-scalar tensor with constant components is not necessarily zero in general coordinate systems? Which term of the covariant derivative of such a tensor will vanish?
    Answer: Because the covariant derivative of a non-scalar tensor is made of sum of terms, and only one of these terms is the ordinary partial derivative of the constant component. Hence, even if this term vanished the other terms (i.e. the Christoffel

---

[27] We note that in the last set of equations we relabeled some dummy indices and used the symmetry of the Christoffel symbols in their paired indices to reach our final result. All these operations should be obvious to the reader. Also, in the last line we used the definition of the Riemann-Christoffel curvature tensor which is given in the book.

symbol terms) do not necessarily vanish and hence the covariant derivative of a non-scalar tensor with constant components is not necessarily zero. In other words, although the partial derivative term will vanish because it represents the rate of change of the constant component, the Christoffel symbol terms which represent the rate of change of the basis vectors do not necessarily vanish because the constancy of the component does not imply the constancy of the basis vectors.

As indicated in the answer of the first part of the question, the vanishing term is the ordinary partial derivative term.

56. Show that: $A_{i;j} = A_{j;i}$ where $\mathbf{A}$ is a gradient of a scalar field.
    Answer: Let f be a differentiable scalar field, $\mathbf{A}$ is its gradient (i.e. $A_i = \partial_i f$) and $A_{i;j}$ is the covariant derivative of this gradient. Hence, we have:
    $$\begin{aligned} A_{i;j} &= \partial_j A_i - \Gamma_{ij}^k A_k \\ &= \partial_j \partial_i f - \Gamma_{ij}^k A_k \\ &= \partial_i \partial_j f - \Gamma_{ij}^k A_k \\ &= \partial_i A_j - \Gamma_{ij}^k A_k \\ &= \partial_i A_j - \Gamma_{ji}^k A_k \\ &= A_{j;i} \end{aligned}$$
    where line 3 is justified by the commutativity of the ordinary partial differential operators, and line 5 is justified by the symmetry of the Christoffel symbols in their paired indices.

57. Show that the covariant derivative of the basis vectors of the covariant and contravariant types is identically zero, i.e. $\mathbf{E}_{i;j} = \mathbf{0}$ and $\mathbf{E}^i_{;j} = \mathbf{0}$.
    Answer: We have:
    $$\begin{aligned} \mathbf{E}_{i;j} &= \partial_j \mathbf{E}_i - \Gamma_{ij}^k \mathbf{E}_k \\ &= \Gamma_{ij}^k \mathbf{E}_k - \Gamma_{ij}^k \mathbf{E}_k \\ &= \mathbf{0} \end{aligned}$$
    where line 2 is based on the identity $\partial_j \mathbf{E}_i = \Gamma_{ij}^k \mathbf{E}_k$ which is given in the book. Similarly:
    $$\begin{aligned} \mathbf{E}^i_{;j} &= \partial_j \mathbf{E}^i + \Gamma_{kj}^i \mathbf{E}^k \\ &= -\Gamma_{kj}^i \mathbf{E}^k + \Gamma_{kj}^i \mathbf{E}^k \\ &= \mathbf{0} \end{aligned}$$
    where line 2 is based on the identity $\partial_j \mathbf{E}^i = -\Gamma_{kj}^i \mathbf{E}^k$ which is given in the book.

58. Prove the following identity:
    $$\partial_k \left(g_{ij} A^i B^j \right) = A_{i;k} B^i + A^i B_{i;k}$$
    Answer: We have:
    $$\partial_k \left(g_{ij} A^i B^j \right) = \left(\partial_k g_{ij} \right) A^i B^j + g_{ij} \left(\partial_k A^i \right) B^j + g_{ij} A^i \left(\partial_k B^j \right)$$

$$
\begin{aligned}
&= \left([ik,j]+[jk,i]\right)A^iB^j + g_{ij}B^j\partial_k A^i + g_{ij}A^i\partial_k B^j \\
&= \left(g_{aj}\Gamma^a_{ik}+g_{ai}\Gamma^a_{jk}\right)A^iB^j + g_{ij}B^j\partial_k A^i + g_{ij}A^i\partial_k B^j \\
&= g_{aj}\Gamma^a_{ik}A^iB^j + g_{ai}\Gamma^a_{jk}A^iB^j + g_{ij}B^j\partial_k A^i + g_{ij}A^i\partial_k B^j \\
&= g_{ij}B^j\partial_k A^i + g_{aj}\Gamma^a_{ik}A^iB^j + g_{ij}A^i\partial_k B^j + g_{ai}\Gamma^a_{jk}A^iB^j \\
&= g_{ij}B^i\partial_k A^j + g_{ij}\Gamma^j_{ak}A^aB^i + g_{ij}A^i\partial_k B^j + g_{ij}\Gamma^j_{ak}A^iB^a \\
&= g_{ij}\left(\partial_k A^j + \Gamma^j_{ak}A^a\right)B^i + A^i g_{ij}\left(\partial_k B^j + \Gamma^j_{ak}B^a\right) \\
&= g_{ij}\left(A^j_{;k}\right)B^i + A^i g_{ij}\left(B^j_{;k}\right) \\
&= A_{i;k}B^i + A^i B_{i;k}
\end{aligned}
$$

where line 1 is the product rule of differentiation, line 2 is the identity $\partial_k g_{ij} = [ik,j] + [jk,i]$, line 3 is the relation between the Christoffel symbols of the first and second kind, line 5 is reordering of terms, line 6 is relabeling of dummy indices with use of symmetry of metric tensor, line 8 is the definition of covariant derivative, and line 9 is an index shifting operation.

59. Define absolute differentiation descriptively and mathematically. What are the other names of absolute derivative?
    Answer: Absolute differentiation of a tensor along a t-parameterized curve $C(t)$ in an nD space with respect to the parameter t is the inner multiplication of the covariant derivative of the tensor and the tangent vector to the curve. The absolute differentiation of a vector $\mathbf{A}$ of covariant type A_i and contravariant type A^i is defined mathematically by:

$$
\begin{aligned}
\frac{\delta A_i}{\delta t} &\equiv A_{i;a}\frac{du^a}{dt} = \frac{dA_i}{dt} - \Gamma^b_{ia}A_b\frac{du^a}{dt} \\
\frac{\delta A^i}{\delta t} &\equiv A^i_{;a}\frac{du^a}{dt} = \frac{dA^i}{dt} + \Gamma^i_{ba}A^b\frac{du^a}{dt}
\end{aligned}
$$

where $\frac{\delta}{\delta t}$ is the absolute differentiation operator, the indexed u are general coordinates and t is the curve parameter while the other symbols are as defined previously. This definition is trivially generalized to tensors of higher ranks.

Absolute derivative is also known as intrinsic derivative or absolute covariant derivative.

60. Write the mathematical expression for the absolute derivative of the tensor field A^{ij}_k which is defined over a space curve $C(t)$.
    Answer:
$$
\frac{\delta A^{ij}_k}{\delta t} = \frac{dA^{ij}_k}{dt} + \Gamma^i_{ba}A^{bj}_k\frac{du^a}{dt} + \Gamma^j_{ba}A^{ib}_k\frac{du^a}{dt} - \Gamma^b_{ka}A^{ij}_b\frac{du^a}{dt}
$$

61. Why the absolute derivative of a differentiable scalar is the same as its ordinary total derivative, i.e. $\frac{\delta f}{\delta t} = \frac{df}{dt}$?
    Answer: Because the covariant derivative of a scalar is the same as the ordinary partial derivative since a scalar has no association with basis vectors to differentiate and therefore there will be no Christoffel symbol terms. Accordingly, the absolute derivative, which is an inner product of the covariant derivative, will have only an

ordinary derivative term and hence it is the same as the ordinary total derivative of the scalar. In symbolic terms:

$$
\begin{aligned}
\frac{\delta f}{\delta t} &= f_{;a}\frac{du^a}{dt} \\
&= f_{,a}\frac{du^a}{dt} \\
&= \frac{\partial f}{\partial u^a}\frac{du^a}{dt} \\
&= \frac{df}{dt}
\end{aligned}
$$

where line 1 is based on the definition of absolute derivative, line 2 is based on the fact that the covariant derivative of a differentiable scalar is the same as its ordinary partial derivative, line 3 is a notation, and line 4 is based on the chain rule in multi-variable differentiation noting that f is parametrically dependent on t only.

62. Why the absolute derivative of a differentiable non-scalar tensor is the same as its ordinary total derivative in rectilinear coordinate systems?

   Answer: Because in rectilinear systems the covariant derivative is the same as the ordinary partial derivative since the Christoffel symbols are zero in these systems. Accordingly, the absolute derivative, which is an inner product of the covariant derivative, will have only an ordinary derivative term and hence it is the same as the ordinary total derivative of the tensor. In symbolic terms (using A^i_{jk} as an instance):

$$
\begin{aligned}
\frac{\delta A^i_{jk}}{\delta t} &= A^i_{jk;a}\frac{du^a}{dt} \\
&= \left(\partial_a A^i_{jk} + 0 - 0 - 0\right)\frac{du^a}{dt} \\
&= \partial_a A^i_{jk}\frac{du^a}{dt} \\
&= \frac{\partial A^i_{jk}}{\partial u^a}\frac{du^a}{dt} \\
&= \frac{dA^i_{jk}}{dt}
\end{aligned}
$$

   where the lines are similarly justified as in the answer of the previous question.

63. From the pattern of covariant derivative of a general tensor, obtain the pattern of its absolute derivative.

   Answer: The pattern of the covariant derivative was explained in detail earlier (see Exercise 34), so all we need to obtain the pattern of absolute derivative is to add the following rule: the covariant derivative is inner multiplied with $\frac{du^a}{dt}$ where the index a is contracted with the differentiation index of the covariant derivative.

64. We have $\mathbf{A} = A^{ij}_{k}\mathbf{E}_i\mathbf{E}_j\mathbf{E}^k$. Apply the ordinary total differentiation process (i.e. $\frac{d\mathbf{A}}{dt}$) onto this tensor (including its basis vectors) to obtain its absolute derivative.

Answer: We have:
$$\begin{aligned}
\frac{d\mathbf{A}}{dt} &= \frac{d}{dt}\left(A^{ij}_{k}\mathbf{E}_i\mathbf{E}_j\mathbf{E}^k\right) \\
&= \left(\frac{dA^{ij}_{k}}{dt}\right)\mathbf{E}_i\mathbf{E}_j\mathbf{E}^k + A^{ij}_{k}\left(\frac{d\mathbf{E}_i}{dt}\right)\mathbf{E}_j\mathbf{E}^k + A^{ij}_{k}\mathbf{E}_i\left(\frac{d\mathbf{E}_j}{dt}\right)\mathbf{E}^k + A^{ij}_{k}\mathbf{E}_i\mathbf{E}_j\left(\frac{d\mathbf{E}^k}{dt}\right) \\
&= \left(\frac{\partial A^{ij}_{k}}{\partial u^a}\frac{du^a}{dt}\right)\mathbf{E}_i\mathbf{E}_j\mathbf{E}^k + A^{ij}_{k}\left(\frac{\partial \mathbf{E}_i}{\partial u^a}\frac{du^a}{dt}\right)\mathbf{E}_j\mathbf{E}^k + \\
&\quad A^{ij}_{k}\mathbf{E}_i\left(\frac{\partial \mathbf{E}_j}{\partial u^a}\frac{du^a}{dt}\right)\mathbf{E}^k + A^{ij}_{k}\mathbf{E}_i\mathbf{E}_j\left(\frac{\partial \mathbf{E}^k}{\partial u^a}\frac{du^a}{dt}\right) \\
&= \left(\frac{\partial A^{ij}_{k}}{\partial u^a}\frac{du^a}{dt}\right)\mathbf{E}_i\mathbf{E}_j\mathbf{E}^k + A^{ij}_{k}\left(\Gamma^b_{ia}\mathbf{E}_b\frac{du^a}{dt}\right)\mathbf{E}_j\mathbf{E}^k + \\
&\quad A^{ij}_{k}\mathbf{E}_i\left(\Gamma^b_{ja}\mathbf{E}_b\frac{du^a}{dt}\right)\mathbf{E}^k + A^{ij}_{k}\mathbf{E}_i\mathbf{E}_j\left(-\Gamma^k_{ba}\mathbf{E}^b\frac{du^a}{dt}\right) \\
&= \left(\frac{\partial A^{ij}_{k}}{\partial u^a}\frac{du^a}{dt}\right)\mathbf{E}_i\mathbf{E}_j\mathbf{E}^k + A^{cj}_{k}\left(\Gamma^i_{ca}\mathbf{E}_i\frac{du^a}{dt}\right)\mathbf{E}_j\mathbf{E}^k + \\
&\quad A^{ic}_{k}\mathbf{E}_i\left(\Gamma^j_{ca}\mathbf{E}_j\frac{du^a}{dt}\right)\mathbf{E}^k + A^{ij}_{c}\mathbf{E}_i\mathbf{E}_j\left(-\Gamma^c_{ka}\mathbf{E}^k\frac{du^a}{dt}\right) \\
&= \left(\frac{\partial A^{ij}_{k}}{\partial u^a}\frac{du^a}{dt} + A^{cj}_{k}\Gamma^i_{ca}\frac{du^a}{dt} + A^{ic}_{k}\Gamma^j_{ca}\frac{du^a}{dt} - A^{ij}_{c}\Gamma^c_{ka}\frac{du^a}{dt}\right)\mathbf{E}_i\mathbf{E}_j\mathbf{E}^k \\
&= \left(\frac{\partial A^{ij}_{k}}{\partial u^a} + A^{cj}_{k}\Gamma^i_{ca} + A^{ic}_{k}\Gamma^j_{ca} - A^{ij}_{c}\Gamma^c_{ka}\right)\frac{du^a}{dt}\mathbf{E}_i\mathbf{E}_j\mathbf{E}^k \\
&= A^{ij}_{k;a}\frac{du^a}{dt}\mathbf{E}_i\mathbf{E}_j\mathbf{E}^k \\
&= \frac{\delta A^{ij}_{k}}{\delta t}\mathbf{E}_i\mathbf{E}_j\mathbf{E}^k
\end{aligned}$$

where equality 2 is the product rule of differentiation, equality 3 is the chain rule of differentiation, equality 4 is the identities $\partial_j\mathbf{E}_i = \Gamma^k_{ij}\mathbf{E}_k$ and $\partial_j\mathbf{E}^i = -\Gamma^i_{kj}\mathbf{E}^k$, equality 5 is relabeling of dummy indices, equality 6 is tidying up, equality 7 is taking common factor, equality 8 is the definition of covariant derivative, and equality 9 is the definition of absolute derivative.

65. Which rules of ordinary total differentiation also apply to intrinsic differentiation and which rules do not? State all these rules symbolically for both ordinary and intrinsic differentiation.

    Answer:
    Linearity: this applies to both ordinary total differentiation and intrinsic differentiation, that is:
    $$\frac{d}{dt}(af + bg) = a\frac{df}{dt} + b\frac{dg}{dt}$$

$$\frac{\delta}{\delta t}\left(a\mathbf{A} \pm b\mathbf{B}\right) = a\frac{\delta \mathbf{A}}{\delta t} \pm b\frac{\delta \mathbf{B}}{\delta t}$$

where a and b are scalar constants, f and g are differentiable scalar functions and $\mathbf{A}$ and $\mathbf{B}$ are differentiable tensors.

The product rule of differentiation: this applies to both ordinary total differentiation and intrinsic differentiation, that is:

$$\frac{d}{dt}(fg) = g\frac{df}{dt} + f\frac{dg}{dt}$$
$$\frac{\delta}{\delta t}(\mathbf{A} \circ \mathbf{B}) = \frac{\delta \mathbf{A}}{\delta t} \circ \mathbf{B} + \mathbf{A} \circ \frac{\delta \mathbf{B}}{\delta t}$$

where the symbol ∘ denotes an inner or outer product operator. However, the order of the tensors in the intrinsic differentiation should be observed.

Commutativity of operators: this applies to ordinary differentiation but not to intrinsic differentiation. In fact, this is no more than the commutativity of ordinary partial operators and non-commutativity of covariant operators. To be more clear, let have a second order covariant differentiation of a vector A_k with respect to the indices i and j. Now, since $A_{k;ji} \neq A_{k;ij}$ then we should have:

$$A_{k;ji}\frac{du^i}{dt} \neq A_{k;ij}\frac{du^j}{dt}$$

So, we have:

$$\partial_i \partial_j = \partial_j \partial_i$$
$$\partial_{;i}\partial_{;j}\frac{du^i}{dt} \neq \partial_{;j}\partial_{;i}\frac{du^j}{dt}$$

where the first equation represents ordinary differentiation (which is partial differentiation because we are assuming the dependency on more than one variable) while the second equation represents the intrinsic differentiation.[28]

66. Using your knowledge about covariant differentiation and the fact that absolute differentiation follows the style of covariant differentiation, obtain all the rules of absolute differentiation of the metric tensor, the Kronecker delta tensor and the index shifting and index replacement operators. Express all these rules in words and in symbols.
Answer: All these rules can be easily obtained from the fact that absolute differentiation is no more than a covariant differentiation followed by an inner product operation.

---

[28] The above explanation is based on a certain interpretation of the property of commutativity with regard to the intrinsic differentiation and the corresponding ordinary differentiation operation. However, we may assume other interpretations and hence the rules and explanation could change. The details are irrelevant to our objective. Moreover, most of these details can be easily obtained from first principles where the basic rule that should be followed is that ordinary differential operators are commutative while tensor differential operators are not. We should also note that we are assuming that intrinsic differentiation works with the second order covariant differentiation as with the first order covariant differentiation.

5   TENSOR DIFFERENTIATION   111

Now, because the covariant differentiation of the metric tensor, the Kronecker delta tensor and the index shifting and index replacement operators is zero, then the absolute differentiation of these should also be zero. These rules can be expressed symbolically as:

$$\frac{\delta \mathbf{g}}{\delta t} = \mathbf{0}$$

$$\frac{\delta \boldsymbol{\delta}}{\delta t} = \mathbf{0}$$

$$\frac{\delta \left(g_{ij} A^j\right)}{\delta t} = g_{ij} \frac{\delta A^j}{\delta t}$$

$$\frac{\delta \left(\delta_i^j A_j\right)}{\delta t} = \delta_i^j \frac{\delta A_j}{\delta t}$$

67. Justify the following statement: "For coordinate systems in which all the components of the metric tensor are constants, the absolute derivative is the same as the ordinary total derivative".
    Answer: Because in such coordinate systems the Christoffel symbols are identically zero (as established earlier in Exercise 20) and hence all the terms of the absolute derivative will vanish except the first which is the ordinary total derivative term. Therefore, the absolute derivative becomes an ordinary total derivative. In symbolic terms (using $A^{ij}_{\ \ k}$ as an instance):

    $$\frac{\delta A^{ij}_{\ \ k}}{\delta t} = \frac{dA^{ij}_{\ \ k}}{dt} + 0 + 0 - 0 = \frac{dA^{ij}_{\ \ k}}{dt}$$

68. The absolute derivative of a tensor along a given curve is unique. What this means?
    Answer: It means that we will obtain the same absolute derivative along the given curve regardless of the coordinate system that we are using. This is based on the objectivity of the absolute derivative since if it is well defined and should have any realistic and useful meaning it should be unique and independent of the employed coordinate system. This is also based on the invariance of tensors (in the general sense of this invariance) across all coordinate systems.

69. Summarize all the main properties and rules that govern tensor differentiation (i.e. covariant and absolute differentiation).
    Answer: The main properties and rules of tensor differentiation are:
    • Tensor differentiation is the same as ordinary differentiation (i.e. partial and total differentiation) but with the application of the differentiation process on both the tensor components and its basis tensor using the product rule of differentiation.
    • The covariant and absolute derivatives of tensors are tensors.
    • The rank of the covariant derivative is 1 covariant rank higher than the rank of the differentiated tensor while the rank of the absolute derivative is the same as the rank of the differentiated tensor. Hence a tensor of type (m, n) will have a covariant derivative of type $(m, n+1)$ and an absolute derivative of type (m, n).[29]

---

[29] We are considering here first order covariant derivative. In brief, each covariant differentiation operation increases the covariant rank by 1.

- The sum and product rules of differentiation apply to tensor differentiation as for ordinary differentiation. However, the order of tensors in tensor differentiation should be respected in the tensor product.
- The covariant and absolute derivatives of scalars and affine tensors of higher ranks are the same as the ordinary derivatives (i.e. partial and total).
- The covariant and absolute derivatives of the metric, Kronecker and permutation tensors as well as the basis vectors vanish identically in any coordinate system.
- Unlike ordinary differential operators, tensor differential operators do not commute with each other.
- Tensor differential operators commute with the contraction of indices.
- Tensor differential operators commute with the index replacement operator and index shifting operators.

Chapter 6
Differential Operations

1. Describe briefly the nabla based differential operators and operations considering the interaction of the nabla operator with the tensors which are acted upon by this operator.
   Answer: In brief:
   • The nabla differential operator may act directly on a tensor (in its general sense that includes scalar and vector) resulting in the gradient of the tensor.
   • The nabla differential operator may act on a non-scalar tensor through dot product multiplication resulting in the divergence of the tensor.
   • The nabla differential operator may act on a non-scalar tensor through cross product multiplication resulting in the curl of the tensor.
   • The nabla differential operator may act on another nabla operator through dot product multiplication resulting in the Laplacian operator.
   There are other less common operations and operators, but they are not investigated in the book.

2. What are the advantages and disadvantages of using the coordinates as suffixes for labeling the operators, basis vectors and tensor components in cylindrical and spherical systems instead of indexed general coordinates? What are the advantages and disadvantages of the opposite?
   Answer: The main advantage of using the coordinates as suffixes is that this notation is intuitive, unambiguous and widely used. The main disadvantage is that it cannot be put in compact tensor form using tensor notation (or indicial notation) which is based on indices.
   The advantage of the opposite is that we can put it in a compact tensor form using tensor notation. The main disadvantage is that it is not as intuitive and commonly used as the use of coordinates; moreover, some ambiguity and confusion may arise with regard to the correspondence between the indices and the coordinates (e.g. if the index 2 in spherical systems refers to θ or ϕ).

3. "The differentiation of a tensor increases its rank by one, by introducing an extra covariant index, unless it implies a contraction in which case it reduces the rank by one". Justify this statement giving common examples from vector and tensor calculus.
   Answer: As seen in the previous chapter, covariant differentiation introduces a new covariant index (i.e. the differentiation index) to the tensor and hence it increases its covariant rank by 1. For example, A_i is a rank-1 tensor but its covariant derivative with respect to the j index is $A_{i;j}$ which is a rank-2 tensor. However, some differential operations include a contraction operation and hence although they introduce a new index they consume two indices by contraction and hence the result is a tensor that is

1 rank lower than the rank of the original tensor.[30]

The most obvious examples are the gradient (which increases the rank of the differentiated tensor by 1 since it introduces 1 differentiation index with no contraction), and the divergence (which reduces the rank of the differentiated tensor by 1 since it introduces 1 differentiation index but consumes 2 indices by contraction).

4. Write the following subsidiary nabla based operators in tensor notation: $\mathbf{A} \cdot \nabla$ and $\mathbf{A} \times \nabla$. Is this notation consistent with the notation of dot and cross product of vectors?

    Answer: Using tensor notation, these operators are defined in Cartesian coordinates as:

    $$\mathbf{A} \cdot \nabla = A_i \partial_i$$
    $$[\mathbf{A} \times \nabla]_i = \epsilon_{ijk} A_j \partial_k$$

    where $\mathbf{A}$ is a vector.

    Yes, this notation is consistent with the notation of dot and cross product of vectors.

5. Why in general we have: $\mathbf{A} \cdot \nabla \neq \nabla \cdot \mathbf{A}$ and $\mathbf{A} \times \nabla \neq \nabla \times \mathbf{A}$?

    Answer: The order is important because it determines the meaning of the operator and the nature of the action that is supposed to be conducted by it. For example, $\mathbf{A} \cdot \nabla$ means that ∇ is not acting on $\mathbf{A}$ but it is acting on something else, while $\nabla \cdot \mathbf{A}$ means that ∇ is acting on $\mathbf{A}$ (i.e. taking the divergence of $\mathbf{A}$) although in both cases $\mathbf{A}$ and ∇ are involved in a dot product operation. Similarly, $\mathbf{A} \times \nabla$ means that ∇ is not acting on $\mathbf{A}$ but it is acting on something else, while $\nabla \times \mathbf{A}$ means that ∇ is acting on $\mathbf{A}$ (i.e. taking the curl of $\mathbf{A}$) although in both cases $\mathbf{A}$ and ∇ are involved in a cross product operation.

6. Define the nabla vector operator and the Laplacian scalar operator in Cartesian coordinate systems using tensor notation.

    Answer:

    $$\nabla_i = \frac{\partial}{\partial x_i} = \partial_i$$
    $$\nabla^2 = \frac{\partial^2}{\partial x_i \partial x_i} = \delta_{ij} \frac{\partial^2}{\partial x_i \partial x_j} = \partial_{ii}$$

7. Find the gradient of the following vector field in a Cartesian coordinate system: $\mathbf{A} = (x, 2x^2, \pi)$.

---

[30] We note that this is different from the absolute differentiation which keeps the type and rank of the original tensor because in absolute differentiation the differentiation index is contracted with the index of the tangent vector and not with an index of the original tensor, i.e. in this process we introduce two indices (one from covariant differentiation and one from the tangent vector) and consume these two indices by contraction without touching the indices of the original tensor and therefore the original tensor keeps its type and rank. This can be easily seen from the definition of absolute differentiation, i.e.

$$\frac{\delta \mathbf{A}}{\delta t} = \mathbf{A}_{;k} \frac{du^k}{dt}$$

Answer: The components of the gradient (which is a rank-2 tensor) are:

$$\partial_x x = 1 \qquad \partial_y x = 0 \qquad \partial_z x = 0$$
$$\partial_x (2x^2) = 4x \qquad \partial_y (2x^2) = 0 \qquad \partial_z (2x^2) = 0$$
$$\partial_x \pi = 0 \qquad \partial_y \pi = 0 \qquad \partial_z \pi = 0$$

Hence, it can be represented by the following matrix:

$$[A_{i,j}] = \begin{bmatrix} 1 & 0 & 0 \\ 4x & 0 & 0 \\ 0 & 0 & 0 \end{bmatrix}$$

where $i,j = 1,2,3$ and i refers to the components of $\mathbf{A}$ while j refers to the coordinates x, y, z.

8. Define the divergence of a differentiable vector descriptively and mathematically assuming a Cartesian coordinate system.

   Answer: The divergence of a differentiable vector field $\mathbf{A}$ is a scalar defined as the dot product of the nabla operator and the vector $\mathbf{A}$ (in this order), that is:

   $$\nabla \cdot \mathbf{A} = \delta_{ij} \frac{\partial A_i}{\partial x_j} = \partial_i A_i$$

   where we are assuming a Cartesian coordinate system.

9. What is the divergence of the following vector field in Cartesian coordinates: $\mathbf{A} = (2z, y^3, e^x)$?

   Answer:

   $$\begin{aligned} \nabla \cdot \mathbf{A} &= \frac{\partial (2z)}{\partial x} + \frac{\partial y^3}{\partial y} + \frac{\partial e^x}{\partial z} \\ &= 0 + 3y^2 + 0 \\ &= 3y^2 \end{aligned}$$

10. Write symbolically, using tensor notation, the following two forms of the divergence of a rank-2 tensor field $\mathbf{A}$ in Cartesian coordinates: $\nabla \cdot \mathbf{A}$ and $\nabla \cdot \mathbf{A}^T$.

    Answer:

    $$\begin{aligned} [\nabla \cdot \mathbf{A}]_i &= \partial_j A_{ji} \\ [\nabla \cdot \mathbf{A}]_j &= \partial_i A_{ji} \end{aligned}$$

11. Define the curl $\nabla \times \mathbf{A}$ in Cartesian coordinates using tensor notation where (a) $\mathbf{A}$ is a rank-1 tensor and (b) $\mathbf{A}$ is a rank-2 tensor (note the two possibilities in the last case).

    Answer:
    (a) $\mathbf{A}$ is a rank-1 tensor:

    $$[\nabla \times \mathbf{A}]_i = \epsilon_{ijk} \partial_j A_k$$

    (b) $\mathbf{A}$ is a rank-2 tensor:

    $$\begin{aligned} {[\nabla \times \mathbf{A}]}_{ij} &= \epsilon_{imn} \partial_m A_{nj} \\ {[\nabla \times \mathbf{A}]}_{ik} &= \epsilon_{imn} \partial_m A_{kn} \end{aligned}$$

12. What is the curl of the following vector field assuming a Cartesian coordinate system: $\mathbf{A} = (5e^{2x}, \pi xy, z^2)$?
    Answer:
    $$\nabla \times \mathbf{A} = \begin{vmatrix} \mathbf{i} & \mathbf{j} & \mathbf{k} \\ \frac{\partial}{\partial x} & \frac{\partial}{\partial y} & \frac{\partial}{\partial z} \\ 5e^{2x} & \pi xy & z^2 \end{vmatrix}$$
    $$= \mathbf{i}\left(\frac{\partial z^2}{\partial y} - \frac{\partial \pi xy}{\partial z}\right) - \mathbf{j}\left(\frac{\partial z^2}{\partial x} - \frac{\partial 5e^{2x}}{\partial z}\right) + \mathbf{k}\left(\frac{\partial \pi xy}{\partial x} - \frac{\partial 5e^{2x}}{\partial y}\right)$$
    $$= \mathbf{i}\,(0-0) - \mathbf{j}\,(0-0) + \mathbf{k}\,(\pi y - 0)$$
    $$= \pi y \mathbf{k}$$

13. Find the Laplacian of the following vector field in Cartesian coordinates:
    $$\mathbf{A} = (x^2 y,\, 2y\sin z,\, \pi z e^{\cosh x})$$

    Answer:
    $$\nabla^2 \mathbf{A} = \nabla^2\left(\mathbf{i}x^2 y + \mathbf{j}2y\sin z + \mathbf{k}\pi z e^{\cosh x}\right)$$
    $$= \mathbf{i}\nabla^2(x^2 y) + \mathbf{j}\nabla^2(2y\sin z) + \mathbf{k}\nabla^2(\pi z e^{\cosh x})$$
    $$= \mathbf{i}\left(\frac{\partial^2 x^2 y}{\partial x^2} + \frac{\partial^2 x^2 y}{\partial y^2} + \frac{\partial^2 x^2 y}{\partial z^2}\right) +$$
    $$\mathbf{j}\left(\frac{\partial^2 2y\sin z}{\partial x^2} + \frac{\partial^2 2y\sin z}{\partial y^2} + \frac{\partial^2 2y\sin z}{\partial z^2}\right) +$$
    $$\mathbf{k}\left(\frac{\partial^2 \pi z e^{\cosh x}}{\partial x^2} + \frac{\partial^2 \pi z e^{\cosh x}}{\partial y^2} + \frac{\partial^2 \pi z e^{\cosh x}}{\partial z^2}\right)$$
    $$= \mathbf{i}\,(2y + 0 + 0) +$$
    $$\mathbf{j}\,(0 + 0 - 2y\sin z) +$$
    $$\mathbf{k}\left(\pi z e^{\cosh x}\sinh^2 x + \pi z e^{\cosh x}\cosh x + 0 + 0\right)$$
    $$= (2y)\mathbf{i} - (2y\sin z)\mathbf{j} + \pi z e^{\cosh x}\left(\sinh^2 x + \cosh x\right)\mathbf{k}$$

14. Define the nabla operator and the Laplacian operator in general coordinate systems using tensor notation.
    Answer:
    $$\nabla = \mathbf{E}^i \partial_i$$
    $$\nabla^2 = \text{div}\,\text{grad} = \nabla \cdot \nabla = \frac{1}{\sqrt{g}}\partial_i\left(\sqrt{g}\, g^{ij}\partial_j\right)$$

    where $\mathbf{E}^i$ is a contravariant basis vector, g is the determinant of the covariant metric tensor and g^{ij} is the contravariant metric tensor.

15. Obtain an expression for the gradient of a covariant vector $\mathbf{A} = A_i \mathbf{E}^i$ in general coordinates justifying each step in your derivation. Repeat the question with a contravariant vector $\mathbf{A} = A^i \mathbf{E}_i$.
    Answer: We have:[31]
    $$\begin{aligned} \nabla \mathbf{A} &= \mathbf{E}^j \partial_j \left(A_i \mathbf{E}^i \right) \\ &= \mathbf{E}^j \mathbf{E}^i \partial_j A_i + \mathbf{E}^j A_i \partial_j \mathbf{E}^i \\ &= \mathbf{E}^j \mathbf{E}^i \partial_j A_i + \mathbf{E}^j A_i \left(-\Gamma^i_{bj} \mathbf{E}^b \right) \\ &= \mathbf{E}^j \mathbf{E}^i \partial_j A_i - \mathbf{E}^j \mathbf{E}^i \Gamma^b_{ij} A_b \\ &= \mathbf{E}^j \mathbf{E}^i \left(\partial_j A_i - \Gamma^b_{ij} A_b \right) \\ &= \mathbf{E}^j \mathbf{E}^i A_{i;j} \end{aligned}$$
    where line 2 is the product rule of differentiation, line 3 is the identity $\partial_j \mathbf{E}^i = -\Gamma^i_{kj} \mathbf{E}^k$, line 4 is relabeling of dummy indices, and line 6 is the definition of covariant derivative. Similarly, we have:
    $$\begin{aligned} \nabla \mathbf{A} &= \mathbf{E}^j \partial_j \left(A^i \mathbf{E}_i \right) \\ &= \mathbf{E}^j \mathbf{E}_i \partial_j A^i + \mathbf{E}^j A^i \partial_j \mathbf{E}_i \\ &= \mathbf{E}^j \mathbf{E}_i \partial_j A^i + \mathbf{E}^j A^i \left(\Gamma^b_{ij} \mathbf{E}_b \right) \\ &= \mathbf{E}^j \mathbf{E}_i \partial_j A^i + \mathbf{E}^j \mathbf{E}_i \Gamma^i_{bj} A^b \\ &= \mathbf{E}^j \mathbf{E}_i \left(\partial_j A^i + \Gamma^i_{bj} A^b \right) \\ &= \mathbf{E}^j \mathbf{E}_i A^i{}_{;j} \end{aligned}$$
    where line 3 is the identity $\partial_j \mathbf{E}_i = \Gamma^k_{ij} \mathbf{E}_k$ while the other lines are justified as in the previous part.
16. Repeat question 15 with a rank-2 mixed tensor $\mathbf{A} = A^i{}_j \mathbf{E}_i \mathbf{E}^j$.
    Answer: We have:
    $$\begin{aligned} \nabla \mathbf{A} &= \mathbf{E}^k \partial_k \left(A^i{}_j \mathbf{E}_i \mathbf{E}^j \right) \\ &= \mathbf{E}^k \left(\partial_k A^i{}_j \right) \mathbf{E}_i \mathbf{E}^j + \mathbf{E}^k A^i{}_j \left(\partial_k \mathbf{E}_i \right) \mathbf{E}^j + \mathbf{E}^k A^i{}_j \mathbf{E}_i \left(\partial_k \mathbf{E}^j \right) \\ &= \mathbf{E}^k \mathbf{E}_i \mathbf{E}^j \partial_k A^i{}_j + \mathbf{E}^k A^i{}_j \left(\Gamma^b_{ik} \mathbf{E}_b \right) \mathbf{E}^j + \mathbf{E}^k A^i{}_j \mathbf{E}_i \left(-\Gamma^j_{bk} \mathbf{E}^b \right) \\ &= \mathbf{E}^k \mathbf{E}_i \mathbf{E}^j \partial_k A^i{}_j + \mathbf{E}^k \mathbf{E}_i \mathbf{E}^j A^b{}_j \Gamma^i_{bk} - \mathbf{E}^k \mathbf{E}_i \mathbf{E}^j A^i{}_b \Gamma^b_{jk} \\ &= \mathbf{E}^k \mathbf{E}_i \mathbf{E}^j \left(\partial_k A^i{}_j + A^b{}_j \Gamma^i_{bk} - A^i{}_b \Gamma^b_{jk} \right) \\ &= \mathbf{E}^k \mathbf{E}_i \mathbf{E}^j A^i{}_{j;k} \end{aligned}$$
    where the lines are justified as in the answer of the previous question.
17. Define, in tensor language, the contravariant form of the gradient of a scalar field f.
    Answer: The contravariant form of the gradient of a scalar field f is given by:
    $$[\nabla f]^i = \partial^i f = g^{ij} \partial_j f = g^{ij} f_{,j} = f^{,i}$$

---

[31] As we noted in the book, the basis vector that associates the derivative operator in the following equations (as well as in similar equations and expressions) should be the last one in the basis tensor.

where g^{ij} is the contravariant metric tensor.

18. Define the divergence of a differentiable vector descriptively and mathematically assuming a general coordinate system.
Answer: The divergence of a differentiable contravariant vector field A^j is a scalar obtained by contracting the differentiation index of the covariant derivative of the vector with the contravariant index of the vector, that is:

$$\nabla \cdot \mathbf{A} = \delta_j^i A_{;i}^j = A_{;i}^i = \frac{1}{\sqrt{g}} \partial_i \left(\sqrt{g} A^i \right)$$

where δ_j^i is the Kronecker delta and g is the determinant of the covariant metric tensor.

19. Derive the following expression for the divergence of a contravariant vector $\mathbf{A}$ in general coordinates: $\nabla \cdot \mathbf{A} = A_{;i}^i$.
Answer: We have:

$$\begin{aligned}
\nabla \cdot \mathbf{A} &= \mathbf{E}^i \partial_i \cdot \left(A^j \mathbf{E}_j \right) \\
&= \mathbf{E}^i \cdot \partial_i \left(A^j \mathbf{E}_j \right) \\
&= \mathbf{E}^i \cdot \left(A_{;i}^j \mathbf{E}_j \right) \\
&= \left(\mathbf{E}^i \cdot \mathbf{E}_j \right) A_{;i}^j \\
&= \delta_j^i A_{;i}^j \\
&= A_{;i}^i
\end{aligned}$$

where line 3 is the definition of covariant derivative, line 5 is the relation between the basis vectors and the mixed type metric tensor, and line 6 is an index replacement operation.

20. Verify the following formula for the divergence of a contravariant vector $\mathbf{A}$ in general coordinates: $\nabla \cdot \mathbf{A} = \frac{1}{\sqrt{g}} \partial_i \left(\sqrt{g} A^i \right)$. Repeat the question with the formula: $\nabla \cdot \mathbf{A} = g^{ji} A_{j;i}$ where $\mathbf{A}$ is a covariant vector.
Answer: We have:

$$\begin{aligned}
\nabla \cdot \mathbf{A} &= A_{;i}^i \\
&= \partial_i A^i + \Gamma_{ji}^i A^j \\
&= \partial_i A^i + A^j \frac{1}{\sqrt{g}} \partial_j \left(\sqrt{g} \right) \\
&= \partial_i A^i + A^i \frac{1}{\sqrt{g}} \partial_i \left(\sqrt{g} \right) \\
&= \frac{1}{\sqrt{g}} \partial_i \left(\sqrt{g} A^i \right)
\end{aligned}$$

where line 1 is the definition of divergence, line 2 is the definition of covariant derivative, line 3 is the identity $\Gamma_{ji}^i = \frac{1}{\sqrt{g}} \partial_j \left(\sqrt{g} \right)$ which was established in the book, line 4 is relabeling the dummy index j, and line 5 is the product rule of differentiation. Similarly, we have:

$$\begin{aligned} g^{ji} A_{j;i} &= \left(g^{ji} A_j\right)_{;i} \\ &= \left(A^i\right)_{;i} \\ &= A^i{}_{;i} \\ &= \nabla \cdot \mathbf{A} \end{aligned}$$

where line 1 is the constancy of the metric tensor with respect to covariant differentiation (Ricci theorem), line 2 is an index raising operation, and line 4 is the definition of the divergence of a contravariant vector.

21. Repeat question 20 with the formula: $\nabla \cdot \mathbf{A} = \mathbf{E}_k A^{ik}{}_{;i}$ where $\mathbf{A}$ is a rank-2 contravariant tensor.
    Answer: We have:

$$\begin{aligned} \nabla \cdot \mathbf{A} &= \mathbf{E}^i \partial_i \cdot \left(A^{jk} \mathbf{E}_j \mathbf{E}_k\right) \\ &= \mathbf{E}^i \cdot \partial_i \left(A^{jk} \mathbf{E}_j \mathbf{E}_k\right) \\ &= \mathbf{E}^i \cdot \left(A^{jk}{}_{;i} \mathbf{E}_j \mathbf{E}_k\right) \\ &= \left(\mathbf{E}^i \cdot \mathbf{E}_j\right) \mathbf{E}_k A^{jk}{}_{;i} \\ &= \delta^i_j \mathbf{E}_k A^{jk}{}_{;i} \\ &= \mathbf{E}_k A^{ik}{}_{;i} \end{aligned}$$

where line 3 is the definition of covariant derivative, line 4 is the intended dot product (since the differentiation index is to be contracted with the first index of the tensor), line 5 is the relation between the basis vectors and the mixed type metric tensor, and line 6 is an index replacement operation.

22. Prove that the divergence of a contravariant vector is a scalar (i.e. rank-0 tensor) by showing that it is invariant under coordinate transformations.
    Answer: As explained earlier, the divergence of a differentiable contravariant vector is the result of contracting the differentiation index of the covariant derivative of the vector with the contravariant index of the vector. Moreover, the covariant derivative of a tensor (in this case a rank-1 contravariant vector) is a tensor which is 1 covariant rank higher than the rank of the original tensor and hence the covariant derivative of a contravariant tensor is a rank-2 mixed type tensor. Now, since the contraction of index of tensors produces a tensor (i.e. invariant under coordinate transformations) which is 2 rank lower than the rank of the original tensor, then the divergence of a contravariant vector is a rank-0 tensor or scalar (i.e. invariant under coordinate transformations), as required.
    There are more formal approaches to this question but the above argument should be sufficient.

23. Derive, from the first principles, the following formula for the curl of a covariant vector field $\mathbf{A}$ in general coordinates: $[\nabla \times \mathbf{A}]^k = \frac{\epsilon^{ijk}}{\sqrt{g}} \left(\partial_i A_j - \Gamma^l_{ji} A_l\right)$.

Answer: We have:

$$
\begin{aligned}
\nabla \times \mathbf{A} &= \mathbf{E}^i \partial_i \times A_j \mathbf{E}^j \\
&= \mathbf{E}^i \times \partial_i \left(A_j \mathbf{E}^j \right) \\
&= \mathbf{E}^i \times \left(A_{j;i} \mathbf{E}^j \right) \\
&= A_{j;i} \left(\mathbf{E}^i \times \mathbf{E}^j \right) \\
&= A_{j;i}\, \underline{\epsilon}^{ijk} \mathbf{E}_k \\
&= \underline{\epsilon}^{ijk} A_{j;i} \mathbf{E}_k \\
&= \frac{\epsilon^{ijk}}{\sqrt{g}} \left(\partial_i A_j - \Gamma^l_{ji} A_l \right) \mathbf{E}_k
\end{aligned}
$$

where line 3 is the definition of covariant derivative, line 5 is an identity about the cross product of basis vectors which was established in the book, and line 7 is the expression of the covariant derivative of a covariant vector plus the definition of the absolute contravariant permutation tensor. Hence, the kth contravariant component of curl $\mathbf{A}$ is:

$$
[\nabla \times \mathbf{A}]^k = \frac{\epsilon^{ijk}}{\sqrt{g}} \left(\partial_i A_j - \Gamma^l_{ji} A_l \right)
$$

24. Show that the formula in exercise 23 will reduce to $[\nabla \times \mathbf{A}]^k = \frac{\epsilon^{ijk}}{\sqrt{g}} \partial_i A_j$ due to the symmetry of the Christoffel symbols in their lower indices.

Answer: The permutation tensor is non-zero only when $i \neq j \neq k$. Hence, for each kth component we have only two non-vanishing terms in the formula of exercise 23. These terms correspond to the two permutations of ij (with $i \neq j \neq k$) where in one of these permutations ϵ^{ijk} is $+1$ and in the other permutation ϵ^{ijk} is -1. Now, if we use upper case indices (i.e. I, J, K) to indicate that these indices have fixed values with $I \neq J \neq K$ (e.g. $I = 1$, $J = 2$ and $K = 3$), then the formula of exercise 23 can be written as:

$$
\begin{aligned}
[\nabla \times \mathbf{A}]^K &= \frac{\epsilon^{IJK}}{\sqrt{g}} \left(\partial_I A_J - \Gamma^b_{JI} A_b \right) + \frac{\epsilon^{JIK}}{\sqrt{g}} \left(\partial_J A_I - \Gamma^b_{IJ} A_b \right) \\
&= \frac{\epsilon^{IJK}}{\sqrt{g}} \left(\partial_I A_J - \Gamma^b_{JI} A_b \right) - \frac{\epsilon^{IJK}}{\sqrt{g}} \left(\partial_J A_I - \Gamma^b_{IJ} A_b \right) \\
&= \frac{\epsilon^{IJK}}{\sqrt{g}} \left(\partial_I A_J - \Gamma^b_{JI} A_b - \partial_J A_I + \Gamma^b_{IJ} A_b \right) \\
&= \frac{\epsilon^{IJK}}{\sqrt{g}} \left(\partial_I A_J - \Gamma^b_{JI} A_b - \partial_J A_I + \Gamma^b_{JI} A_b \right) \\
&= \frac{\epsilon^{IJK}}{\sqrt{g}} \left(\partial_I A_J - \partial_J A_I \right) \\
&= \frac{\epsilon^{IJK}}{\sqrt{g}} \partial_I A_J + \frac{\epsilon^{JIK}}{\sqrt{g}} \partial_J A_I
\end{aligned}
$$

where line 2 is based on the fact that the permutation tensor is totally anti-symmetric, and line 4 is based on the symmetry of the Christoffel symbols of the second kind in their lower indices. Now, if we return to our ordinary lower case index notation then the last line is no more than a sum of terms representing all the permutations of ijk (including the zero terms which correspond to the repetitive permutations), and hence the last equation can be written compactly as:

$$[\nabla \times \mathbf{A}]^k = \frac{\epsilon^{ijk}}{\sqrt{g}} \partial_i A_j$$

25. Derive, from the first principles, the following expression for the Laplacian of a scalar field f in general coordinates: $\nabla^2 f = \frac{1}{\sqrt{g}} \partial_i \left(\sqrt{g} g^{ij} \partial_j f \right)$.
    Answer:
    $$\begin{aligned}
    \nabla^2 f &= \nabla \cdot (\nabla f) \\
    &= \mathbf{E}^i \partial_i \cdot \left(\mathbf{E}^j \partial_j f \right) \\
    &= \mathbf{E}^i \cdot \partial_i \left(\mathbf{E}^j \partial_j f \right) \\
    &= \mathbf{E}^i \cdot \partial_i \left(\mathbf{E}^j f_{,j} \right) \\
    &= \mathbf{E}^i \cdot \left(\mathbf{E}^j f_{,j;i} \right) \\
    &= \left(\mathbf{E}^i \cdot \mathbf{E}^j \right) f_{,j;i} \\
    &= g^{ij} f_{,j;i} \\
    &= \left(g^{ij} f_{,j} \right)_{;i} \\
    &= \left(g^{ij} \partial_j f \right)_{;i} \\
    &= \partial_i \left(g^{ij} \partial_j f \right) + \left(g^{kj} \partial_j f \right) \Gamma^i_{ki} \\
    &= \partial_i \left(g^{ij} \partial_j f \right) + \left(g^{ij} \partial_j f \right) \Gamma^k_{ik} \\
    &= \partial_i \left(g^{ij} \partial_j f \right) + \left(g^{ij} \partial_j f \right) \frac{1}{\sqrt{g}} \left(\partial_i \sqrt{g} \right) \\
    &= \frac{1}{\sqrt{g}} \left[\sqrt{g} \partial_i \left(g^{ij} \partial_j f \right) + \left(g^{ij} \partial_j f \right) \partial_i \sqrt{g} \right] \\
    &= \frac{1}{\sqrt{g}} \partial_i \left(\sqrt{g} g^{ij} \partial_j f \right)
    \end{aligned}$$
    where line 1 is the definition of Laplacian as divergence of gradient, line 5 is the definition of covariant derivative, line 8 is the constancy of the metric tensor with respect to tensor differentiation (Ricci theorem), line 10 is the expression of the covariant derivative of a contravariant vector (i.e. $g^{ij} \partial_j f = \partial^i f$), line 12 is the identity $\Gamma^i_{ji} = \frac{1}{\sqrt{g}} \partial_j \left(\sqrt{g} \right)$ which was established in the book, and line 14 is the product rule of differentiation while the other lines are obvious or justified earlier.

26. Why the basic definition of the Laplacian of a scalar field f in general coordinates as $\nabla^2 f = \text{div}\,(\text{grad}\,f)$ cannot be used as it is to develop a formula before raising the index of the gradient?

Answer: Because the divergence in the above definition of Laplacian implies a contraction operation between the gradient index and the divergence index, and since the contraction operation in general coordinate systems should be between a covariant index and a contravariant index then the index of the gradient (which is a covariant index) should be raised before the contraction operation can take place.

27. Define, in tensor language, the nabla operator and the Laplacian operator assuming an orthogonal coordinate system of a 3D space.
Answer: They are:[32]

$$\nabla = \sum_i \frac{\mathbf{q}_i}{h_i} \frac{\partial}{\partial q^i}$$

$$\nabla^2 = \frac{1}{h_1 h_2 h_3} \sum_{i=1}^{3} \frac{\partial}{\partial q^i} \left(\frac{h_1 h_2 h_3}{(h_i)^2} \frac{\partial}{\partial q^i} \right)$$

where $\mathbf{q}_i$ are basis unit vectors of orthogonal systems, q^i are general orthogonal coordinates and the indexed h are scale factors.

28. Using the expression of the divergence of a vector field in general coordinates, obtain an expression for the divergence in orthogonal coordinates of a 3D space.
Answer: We have:

$$\begin{aligned}
\nabla \cdot \mathbf{A} &= \frac{1}{\sqrt{g}} \partial_i \left(\sqrt{g} A^i \right) \\
&= \frac{1}{\sqrt{g}} \frac{\partial}{\partial q^i} \left(\sqrt{g} A^i \right) \\
&= \frac{1}{\sqrt{g}} \sum_{i=1}^{3} \frac{\partial}{\partial q^i} \left(\sqrt{g} A^i \right) \\
&= \frac{1}{h_1 h_2 h_3} \sum_{i=1}^{3} \frac{\partial}{\partial q^i} \left(h_1 h_2 h_3 A^i \right) \\
&= \frac{1}{h_1 h_2 h_3} \sum_{i=1}^{3} \frac{\partial}{\partial q^i} \left(\frac{h_1 h_2 h_3}{h_i} \hat{A}^i \right) \\
&= \frac{1}{h_1 h_2 h_3} \left[\frac{\partial}{\partial q^1} \left(h_2 h_3 \hat{A}_1 \right) + \frac{\partial}{\partial q^2} \left(h_1 h_3 \hat{A}_2 \right) + \frac{\partial}{\partial q^3} \left(h_1 h_2 \hat{A}_3 \right) \right]
\end{aligned}$$

where line 1 is the expression of the divergence in general coordinates which we obtained earlier (see Exercise 20), line 2 is based on the fact that the system is orthogonal and hence we use orthogonal coordinates, line 3 is based on the summation convention and assuming a 3D space, line 4 is based on the fact that in orthogonal systems the metric tensor is diagonal and hence g (which is the determinant of the covariant metric tensor)

---

[32] In fact, these are vector calculus definitions more than tensor calculus definitions. However, we prefer these for clarity and to avoid lengthy clarifications.

is given by:
$$g = g_{11}g_{22}g_{33} = (h_1 h_2 h_3)^2$$
and line 5 is based on using physical components $\hat{A}^i$ ($= h_i A^i$ with no sum on i) while line 6 is just an expansion of line 5 for the sake of clarity.

29. Define the curl of a vector field **A** in orthogonal coordinates of a 3D space using determinantal form and tensor notation form.
   Answer:
   Determinantal form:
   $$\nabla \times \mathbf{A} = \frac{1}{h_1 h_2 h_3} \begin{vmatrix} h_1 \mathbf{q}_1 & h_2 \mathbf{q}_2 & h_3 \mathbf{q}_3 \\ \frac{\partial}{\partial q^1} & \frac{\partial}{\partial q^2} & \frac{\partial}{\partial q^3} \\ h_1 \hat{A}_1 & h_2 \hat{A}_2 & h_3 \hat{A}_3 \end{vmatrix}$$
   where the symbols are as explained in the last two questions.
   Tensor notation form:
   $$[\nabla \times \mathbf{A}]_i = \sum_{k=1}^{3} \frac{\epsilon_{ijk} h_i}{h_1 h_2 h_3} \frac{\partial (h_k \hat{A}_k)}{\partial q^j}$$
   with no sum over i.

30. Using the expression of the Laplacian of a scalar field in general coordinates, derive an expression for the Laplacian in orthogonal coordinates of a 3D space.
   Answer: We have:
   $$\begin{aligned}
   \nabla^2 f &= \frac{1}{\sqrt{g}} \partial_i \left(\sqrt{g} g^{ij} \partial_j f \right) \\
   &= \frac{1}{\sqrt{g}} \frac{\partial}{\partial q^i} \left(\sqrt{g} g^{ij} \frac{\partial f}{\partial q^j} \right) \\
   &= \frac{1}{\sqrt{g}} \sum_{i=1}^{3} \frac{\partial}{\partial q^i} \left(\sqrt{g} \sum_{j=1}^{3} g^{ij} \frac{\partial f}{\partial q^j} \right) \\
   &= \frac{1}{\sqrt{g}} \sum_{i=1}^{3} \frac{\partial}{\partial q^i} \left(\sqrt{g} g^{ii} \frac{\partial f}{\partial q^i} \right) \\
   &= \frac{1}{h_1 h_2 h_3} \sum_{i=1}^{3} \frac{\partial}{\partial q^i} \left(\frac{h_1 h_2 h_3}{(h_i)^2} \frac{\partial f}{\partial q^i} \right)
   \end{aligned}$$
   where line 1 is the expression of the Laplacian in general coordinates which we obtained earlier (see Exercise 25), line 2 is based on the fact that the system is orthogonal and hence we use orthogonal coordinates, line 3 is based on the summation convention and assuming a 3D space, line 4 is based on the fact that in orthogonal systems we have $g^{ij} = 0$ when $i \neq j$ and hence the sum over j reduces to $g^{ii} \frac{\partial f}{\partial q^i}$ where g^{ii} represents the ith diagonal component, and line 5 is based on the fact that in orthogonal systems we have:
   $$\sqrt{g} = h_1 h_2 h_3 \qquad \text{and} \qquad g^{ii} = \frac{1}{(h_i)^2}$$

6 DIFFERENTIAL OPERATIONS 124

31. Why the components of tensors in cylindrical and spherical coordinates are physical?
    Answer: Because all the basis vectors are normalized and hence they are dimensionless with unit magnitude.[33] Consequently, all the components in these systems are physical with unified physical dimensions.

32. Define the nabla and Laplacian operators in cylindrical coordinates.
    Answer: They are:
    $$\nabla = \mathbf{e}_\rho \partial_\rho + \mathbf{e}_\phi \frac{1}{\rho} \partial_\phi + \mathbf{e}_z \partial_z$$
    $$\nabla^2 = \partial_{\rho\rho} + \frac{1}{\rho} \partial_\rho + \frac{1}{\rho^2} \partial_{\phi\phi} + \partial_{zz}$$
    where $\mathbf{e}_\rho, \mathbf{e}_\phi, \mathbf{e}_z$ are unit basis vectors and $\partial_{\rho\rho} = \partial_\rho \partial_\rho$, $\partial_{\phi\phi} = \partial_\phi \partial_\phi$ and $\partial_{zz} = \partial_z \partial_z$.

33. Use the definition of the gradient of a scalar field f in orthogonal coordinates and the table of scale factors (which is given in the book) to obtain an expression for the gradient in cylindrical coordinates.
    Answer: The gradient of a differentiable scalar field f in orthogonal coordinate systems of a 3D space is given by:
    $$\nabla f = \frac{\mathbf{q}_1}{h_1} \frac{\partial f}{\partial q^1} + \frac{\mathbf{q}_2}{h_2} \frac{\partial f}{\partial q^2} + \frac{\mathbf{q}_3}{h_3} \frac{\partial f}{\partial q^3}$$
    Now, in cylindrical systems we have:
    $$q^1 = \rho \qquad q^2 = \phi \qquad q^3 = z$$
    $$\mathbf{q}_1 = \mathbf{e}_\rho \qquad \mathbf{q}_2 = \mathbf{e}_\phi \qquad \mathbf{q}_3 = \mathbf{e}_z$$
    $$h_1 = 1 \qquad h_2 = \rho \qquad h_3 = 1$$
    Hence, the above equation becomes:
    $$\nabla f = \mathbf{e}_\rho \frac{\partial f}{\partial \rho} + \mathbf{e}_\phi \frac{1}{\rho} \frac{\partial f}{\partial \phi} + \mathbf{e}_z \frac{\partial f}{\partial z}$$

34. Use the definition of the divergence of a vector field $\mathbf{A}$ in orthogonal coordinates and the table of scale factors (which is given in the book) to obtain an expression for the divergence in cylindrical coordinates.
    Answer: The divergence of a differentiable vector field $\mathbf{A}$ in orthogonal coordinate systems of a 3D space is given by:
    $$\nabla \cdot \mathbf{A} = \frac{1}{h_1 h_2 h_3} \left[\frac{\partial}{\partial q^1} \left(h_2 h_3 \hat{A}_1 \right) + \frac{\partial}{\partial q^2} \left(h_1 h_3 \hat{A}_2 \right) + \frac{\partial}{\partial q^3} \left(h_1 h_2 \hat{A}_3 \right) \right]$$
    Now, in cylindrical systems we have:

---
[33] In this question and answer we are referring to the commonly used cylindrical and spherical coordinate systems noting that these systems can also be represented by covariant and contravariant forms.

6 *DIFFERENTIAL OPERATIONS* 125

$$q^1 = \rho \qquad q^2 = \phi \qquad q^3 = z$$
$$h_1 = 1 \qquad h_2 = \rho \qquad h_3 = 1$$
$$\hat{A}_1 = A_\rho \qquad \hat{A}_2 = A_\phi \qquad \hat{A}_3 = A_z$$

Hence, the above equation becomes:

$$\nabla \cdot \mathbf{A} = \frac{1}{\rho} \left[\frac{\partial}{\partial \rho} (\rho A_\rho) + \frac{\partial}{\partial \phi} (A_\phi) + \frac{\partial}{\partial z} (\rho A_z) \right]$$
$$= \frac{1}{\rho} \left[\frac{\partial}{\partial \rho} (\rho A_\rho) + \frac{\partial A_\phi}{\partial \phi} + \rho \frac{\partial A_z}{\partial z} \right]$$

35. Write the determinantal form of the curl of a vector field **A** in cylindrical coordinates.
    Answer:

$$\nabla \times \mathbf{A} = \frac{1}{\rho} \begin{vmatrix} \mathbf{e}_\rho & \rho \mathbf{e}_\phi & \mathbf{e}_z \\ \partial_\rho & \partial_\phi & \partial_z \\ A_\rho & \rho A_\phi & A_z \end{vmatrix}$$

where the symbols are as explained earlier.

36. Use the definition of the Laplacian of a scalar field f in orthogonal coordinates and the table of scale factors (which is given in the book) to obtain an expression for the Laplacian in cylindrical coordinates.
    Answer: The Laplacian of a differentiable scalar field f in orthogonal coordinate systems of a 3D space is given by:

$$\nabla^2 f = \frac{1}{h_1 h_2 h_3} \sum_{i=1}^{3} \frac{\partial}{\partial q^i} \left(\frac{h_1 h_2 h_3}{(h_i)^2} \frac{\partial f}{\partial q^i} \right)$$

Now, in cylindrical systems we have:

$$q^1 = \rho \qquad q^2 = \phi \qquad q^3 = z$$
$$h_1 = 1 \qquad h_2 = \rho \qquad h_3 = 1$$

Hence, the above equation becomes:

$$\nabla^2 f = \frac{1}{\rho} \frac{\partial}{\partial \rho} \left(\rho \frac{\partial f}{\partial \rho} \right) + \frac{1}{\rho} \frac{\partial}{\partial \phi} \left(\frac{1}{\rho} \frac{\partial f}{\partial \phi} \right) + \frac{1}{\rho} \frac{\partial}{\partial z} \left(\rho \frac{\partial f}{\partial z} \right)$$
$$= \frac{\rho}{\rho} \frac{\partial}{\partial \rho} \left(\frac{\partial f}{\partial \rho} \right) + \frac{1}{\rho} \frac{\partial \rho}{\partial \rho} \frac{\partial f}{\partial \rho} + \frac{1}{\rho} \frac{\partial}{\partial \phi} \left(\frac{1}{\rho} \frac{\partial f}{\partial \phi} \right) + \frac{1}{\rho} \frac{\partial}{\partial z} \left(\rho \frac{\partial f}{\partial z} \right)$$
$$= \frac{\partial^2 f}{\partial \rho^2} + \frac{1}{\rho} \frac{\partial f}{\partial \rho} + \frac{1}{\rho} \frac{\partial}{\partial \phi} \left(\frac{1}{\rho} \frac{\partial f}{\partial \phi} \right) + \frac{1}{\rho} \frac{\partial}{\partial z} \left(\rho \frac{\partial f}{\partial z} \right)$$
$$= \frac{\partial^2 f}{\partial \rho^2} + \frac{1}{\rho} \frac{\partial f}{\partial \rho} + \frac{1}{\rho} \frac{1}{\rho} \frac{\partial}{\partial \phi} \left(\frac{\partial f}{\partial \phi} \right) + \frac{\rho}{\rho} \frac{\partial}{\partial z} \left(\frac{\partial f}{\partial z} \right)$$
$$= \frac{\partial^2 f}{\partial \rho^2} + \frac{1}{\rho} \frac{\partial f}{\partial \rho} + \frac{1}{\rho^2} \frac{\partial^2 f}{\partial \phi^2} + \frac{\partial^2 f}{\partial z^2}$$

$$= \partial_{\rho\rho}f + \frac{1}{\rho}\partial_\rho f + \frac{1}{\rho^2}\partial_{\phi\phi}f + \partial_{zz}f$$

where in line 2 we used the product rule of differentiation, and in line 4 we used the fact that the coordinates are independent of each other.

37. A scalar field in cylindrical coordinates is given by: $f(\rho,\phi,z) = \rho$. What are the gradient and Laplacian of this field?
Answer: We have:

$$\begin{aligned}
\nabla f &= \mathbf{e}_\rho \partial_\rho f + \mathbf{e}_\phi \frac{1}{\rho}\partial_\phi f + \mathbf{e}_z \partial_z f \\
&= \mathbf{e}_\rho \partial_\rho \rho + \mathbf{e}_\phi \frac{1}{\rho}\partial_\phi \rho + \mathbf{e}_z \partial_z \rho \\
&= \mathbf{e}_\rho + 0 + 0 \\
&= \mathbf{e}_\rho
\end{aligned}$$

$$\begin{aligned}
\nabla^2 f &= \partial_{\rho\rho}f + \frac{1}{\rho}\partial_\rho f + \frac{1}{\rho^2}\partial_{\phi\phi}f + \partial_{zz}f \\
&= \partial_{\rho\rho}\rho + \frac{1}{\rho}\partial_\rho \rho + \frac{1}{\rho^2}\partial_{\phi\phi}\rho + \partial_{zz}\rho \\
&= 0 + \frac{1}{\rho} + 0 + 0 \\
&= \frac{1}{\rho}
\end{aligned}$$

38. A vector field in cylindrical coordinates is given by: $\mathbf{A}(\rho,\phi,z) = (3z, \pi\phi^2, z^2 \cos\rho)$. What are the divergence and curl of this field?
Answer: We have:

$$\begin{aligned}
\nabla \cdot \mathbf{A} &= \frac{1}{\rho}\left[\partial_\rho(\rho A_\rho) + \partial_\phi A_\phi + \rho\partial_z A_z\right] \\
&= \frac{1}{\rho}\left[\partial_\rho(\rho 3z) + \partial_\phi(\pi\phi^2) + \rho\partial_z(z^2 \cos\rho)\right] \\
&= \frac{1}{\rho}\left[3z + 2\pi\phi + 2\rho z \cos\rho\right]
\end{aligned}$$

$$\begin{aligned}
\nabla \times \mathbf{A} &= \frac{1}{\rho}\begin{vmatrix} \mathbf{e}_\rho & \rho\mathbf{e}_\phi & \mathbf{e}_z \\ \partial_\rho & \partial_\phi & \partial_z \\ A_\rho & \rho A_\phi & A_z \end{vmatrix} \\
&= \frac{1}{\rho}\begin{vmatrix} \mathbf{e}_\rho & \rho\mathbf{e}_\phi & \mathbf{e}_z \\ \partial_\rho & \partial_\phi & \partial_z \\ 3z & \rho\pi\phi^2 & z^2 \cos\rho \end{vmatrix} \\
&= \frac{1}{\rho}\mathbf{e}_\rho\left[\partial_\phi(z^2 \cos\rho) - \partial_z(\rho\pi\phi^2)\right] -
\end{aligned}$$

$$\frac{1}{\rho}\rho\mathbf{e}_\phi\left[\partial_\rho\left(z^2\cos\rho\right)-\partial_z(3z)\right]+$$
$$\frac{1}{\rho}\mathbf{e}_z\left[\partial_\rho\left(\rho\pi\phi^2\right)-\partial_\phi(3z)\right]$$
$$=\frac{1}{\rho}\mathbf{e}_\rho[0-0]-\mathbf{e}_\phi\left[-z^2\sin\rho-3\right]+\frac{1}{\rho}\mathbf{e}_z\left[\pi\phi^2-0\right]$$
$$=\mathbf{e}_\phi\left(z^2\sin\rho+3\right)+\mathbf{e}_z\left(\frac{\pi}{\rho}\phi^2\right)$$

39. Repeat exercise 33 with spherical coordinates.
    Answer: The gradient of a differentiable scalar field f in orthogonal coordinate systems of a 3D space is given by:
    $$\nabla f = \frac{\mathbf{q}_1}{h_1}\frac{\partial f}{\partial q^1}+\frac{\mathbf{q}_2}{h_2}\frac{\partial f}{\partial q^2}+\frac{\mathbf{q}_3}{h_3}\frac{\partial f}{\partial q^3}$$
    Now, in spherical systems we have:

    $q^1 = r$ $\qquad q^2 = \theta \qquad$ $q^3 = \phi$
    $\mathbf{q}_1 = \mathbf{e}_r$ $\qquad \mathbf{q}_2 = \mathbf{e}_\theta \qquad$ $\mathbf{q}_3 = \mathbf{e}_\phi$
    $h_1 = 1$ $\qquad h_2 = r \qquad$ $h_3 = r\sin\theta$

    Hence, the above equation becomes:
    $$\nabla f = \mathbf{e}_r\frac{\partial f}{\partial r}+\frac{\mathbf{e}_\theta}{r}\frac{\partial f}{\partial \theta}+\frac{\mathbf{e}_\phi}{r\sin\theta}\frac{\partial f}{\partial \phi}$$

40. Repeat exercise 34 with spherical coordinates.
    Answer: The divergence of a differentiable vector field $\mathbf{A}$ in orthogonal coordinate systems of a 3D space is given by:
    $$\nabla\cdot\mathbf{A} = \frac{1}{h_1 h_2 h_3}\left[\frac{\partial}{\partial q^1}\left(h_2 h_3 \hat{A}_1\right)+\frac{\partial}{\partial q^2}\left(h_1 h_3 \hat{A}_2\right)+\frac{\partial}{\partial q^3}\left(h_1 h_2 \hat{A}_3\right)\right]$$
    Now, in spherical systems we have:

    $q^1 = r$ $\qquad q^2 = \theta \qquad$ $q^3 = \phi$
    $h_1 = 1$ $\qquad h_2 = r \qquad$ $h_3 = r\sin\theta$
    $\hat{A}_1 = A_r$ $\qquad \hat{A}_2 = A_\theta \qquad$ $\hat{A}_3 = A_\phi$

    Hence, the above equation becomes:
    $$\nabla\cdot\mathbf{A} = \frac{1}{r^2\sin\theta}\left[\frac{\partial}{\partial r}\left(r^2\sin\theta A_r\right)+\frac{\partial}{\partial \theta}\left(r\sin\theta A_\theta\right)+\frac{\partial}{\partial \phi}\left(rA_\phi\right)\right]$$
    $$= \frac{1}{r^2\sin\theta}\left[\sin\theta\frac{\partial\left(r^2 A_r\right)}{\partial r}+r\frac{\partial\left(\sin\theta A_\theta\right)}{\partial \theta}+r\frac{\partial A_\phi}{\partial \phi}\right]$$

41. Repeat exercise 35 with spherical coordinates.
    Answer:
    $$\nabla \times \mathbf{A} = \frac{1}{r^2 \sin\theta} \begin{vmatrix} \mathbf{e}_r & r\mathbf{e}_\theta & r\sin\theta \mathbf{e}_\phi \\ \partial_r & \partial_\theta & \partial_\phi \\ A_r & rA_\theta & r\sin\theta A_\phi \end{vmatrix}$$
    where the symbols are as explained earlier.

42. Repeat exercise 36 with spherical coordinates.
    Answer: The Laplacian of a differentiable scalar field f in orthogonal coordinate systems of a 3D space is given by:
    $$\nabla^2 f = \frac{1}{h_1 h_2 h_3} \sum_{i=1}^{3} \frac{\partial}{\partial q^i}\left(\frac{h_1 h_2 h_3}{(h_i)^2} \frac{\partial f}{\partial q^i}\right)$$

    Now, in spherical systems we have:
    $$q^1 = r \qquad q^2 = \theta \qquad q^3 = \phi$$
    $$h_1 = 1 \qquad h_2 = r \qquad h_3 = r\sin\theta$$

    Hence, the above equation becomes:
    $$\begin{aligned}\nabla^2 f &= \frac{1}{r^2 \sin\theta} \frac{\partial}{\partial r}\left(r^2 \sin\theta \frac{\partial f}{\partial r}\right) + \frac{1}{r^2 \sin\theta} \frac{\partial}{\partial \theta}\left(\frac{r^2 \sin\theta}{r^2} \frac{\partial f}{\partial \theta}\right) + \\ &\quad \frac{1}{r^2 \sin\theta} \frac{\partial}{\partial \phi}\left(\frac{r^2 \sin\theta}{r^2 \sin^2\theta} \frac{\partial f}{\partial \phi}\right) \\ &= \frac{1}{r^2} \frac{\partial}{\partial r}\left(r^2 \frac{\partial f}{\partial r}\right) + \frac{1}{r^2 \sin\theta} \frac{\partial}{\partial \theta}\left(\sin\theta \frac{\partial f}{\partial \theta}\right) + \frac{1}{r^2 \sin^2\theta} \frac{\partial^2 f}{\partial \phi^2} \\ &= \frac{r^2}{r^2} \frac{\partial^2 f}{\partial r^2} + \frac{1}{r^2} \frac{\partial r^2}{\partial r} \frac{\partial f}{\partial r} + \frac{\sin\theta}{r^2 \sin\theta} \frac{\partial^2 f}{\partial \theta^2} + \frac{\cos\theta}{r^2 \sin\theta} \frac{\partial f}{\partial \theta} + \frac{1}{r^2 \sin^2\theta} \frac{\partial^2 f}{\partial \phi^2} \\ &= \frac{\partial^2 f}{\partial r^2} + \frac{2}{r} \frac{\partial f}{\partial r} + \frac{1}{r^2} \frac{\partial^2 f}{\partial \theta^2} + \frac{\cos\theta}{r^2 \sin\theta} \frac{\partial f}{\partial \theta} + \frac{1}{r^2 \sin^2\theta} \frac{\partial^2 f}{\partial \phi^2}\end{aligned}$$

    where in step 3 we used the product rule of differentiation.

43. A scalar field in spherical coordinates is given by: $f(r, \theta, \phi) = r^2 + \theta$. What are the gradient and Laplacian of this field?
    Answer: We have:
    $$\begin{aligned}\nabla f &= \mathbf{e}_r \partial_r f + \mathbf{e}_\theta \frac{1}{r} \partial_\theta f + \mathbf{e}_\phi \frac{1}{r\sin\theta} \partial_\phi f \\ &= \mathbf{e}_r 2r + \mathbf{e}_\theta \frac{1}{r} + 0 \\ &= \mathbf{e}_r 2r + \mathbf{e}_\theta \frac{1}{r}\end{aligned}$$

    $$\nabla^2 f = \partial_{rr} f + \frac{2}{r} \partial_r f + \frac{1}{r^2} \partial_{\theta\theta} f + \frac{\cos\theta}{r^2 \sin\theta} \partial_\theta f + \frac{1}{r^2 \sin^2\theta} \partial_{\phi\phi} f$$

6 DIFFERENTIAL OPERATIONS

$$= 2 + \frac{2}{r}(2r) + 0 + \frac{\cos\theta}{r^2 \sin\theta} + 0$$

$$= 6 + \frac{\cos\theta}{r^2 \sin\theta}$$

44. A vector field in spherical coordinates is given by: $\mathbf{A}(r, \theta, \phi) = (e^r, 5\sin\phi, \ln\theta)$. What are the divergence and curl of this field?
Answer: We have:

$$\nabla \cdot \mathbf{A} = \frac{1}{r^2 \sin\theta} \left[\sin\theta \frac{\partial (r^2 A_r)}{\partial r} + r \frac{\partial (\sin\theta A_\theta)}{\partial \theta} + r \frac{\partial A_\phi}{\partial \phi} \right]$$

$$= \frac{1}{r^2 \sin\theta} \left[\sin\theta \frac{\partial (r^2 e^r)}{\partial r} + r \frac{\partial (\sin\theta\, 5\sin\phi)}{\partial \theta} + r \frac{\partial \ln\theta}{\partial \phi} \right]$$

$$= \frac{1}{r^2 \sin\theta} \left[\sin\theta (2re^r) + \sin\theta (r^2 e^r) + 5r\cos\theta \sin\phi + 0 \right]$$

$$= \frac{1}{r^2 \sin\theta} \left[2re^r \sin\theta + r^2 e^r \sin\theta + 5r\cos\theta \sin\phi \right]$$

$$= \frac{1}{r \sin\theta} \left[2e^r \sin\theta + re^r \sin\theta + 5\cos\theta \sin\phi \right]$$

where in line 3 we used the product rule of differentiation.

$$\nabla \times \mathbf{A} = \frac{1}{r^2 \sin\theta} \begin{vmatrix} \mathbf{e}_r & r\mathbf{e}_\theta & r\sin\theta\,\mathbf{e}_\phi \\ \partial_r & \partial_\theta & \partial_\phi \\ A_r & rA_\theta & r\sin\theta A_\phi \end{vmatrix}$$

$$= \frac{1}{r^2 \sin\theta} \begin{vmatrix} \mathbf{e}_r & r\mathbf{e}_\theta & r\sin\theta\,\mathbf{e}_\phi \\ \partial_r & \partial_\theta & \partial_\phi \\ e^r & r5\sin\phi & r\sin\theta \ln\theta \end{vmatrix}$$

$$= \frac{\mathbf{e}_r}{r^2 \sin\theta} \left[\partial_\theta (r\sin\theta \ln\theta) - \partial_\phi (r5\sin\phi) \right] -$$
$$\frac{r\mathbf{e}_\theta}{r^2 \sin\theta} \left[\partial_r (r\sin\theta \ln\theta) - \partial_\phi (e^r) \right] +$$
$$\frac{r\sin\theta\,\mathbf{e}_\phi}{r^2 \sin\theta} \left[\partial_r (r5\sin\phi) - \partial_\theta (e^r) \right]$$

$$= \frac{\mathbf{e}_r}{r^2 \sin\theta} \left[r\cos\theta \ln\theta + r\sin\theta \frac{1}{\theta} - r5\cos\phi \right] -$$
$$\frac{\mathbf{e}_\theta}{r \sin\theta} \left[\sin\theta \ln\theta - 0 \right] +$$
$$\frac{\mathbf{e}_\phi}{r} \left[5\sin\phi - 0 \right]$$

$$= \frac{\mathbf{e}_r}{r\sin\theta} \left(\cos\theta \ln\theta + \frac{1}{\theta}\sin\theta - 5\cos\phi \right) - \mathbf{e}_\theta \frac{\ln\theta}{r} + \mathbf{e}_\phi \frac{5\sin\phi}{r}$$

Chapter 7
Tensors in Application

1. Summarize the reasons for the popularity of tensor calculus techniques in mathematical, scientific and engineering applications.
   Answer: These techniques are beautiful, powerful and succinct.
2. State, in tensor language, the definition of the following mathematical concepts assuming Cartesian coordinates of a 3D space: trace of matrix, determinant of matrix, inverse of matrix, multiplication of two compatible square matrices, dot product of two vectors, cross product of two vectors, scalar triple product of three vectors and vector triple product of three vectors.
   Answer:

$$\text{tr}(\mathbf{A}) = A_{ii}$$
$$\det(\mathbf{A}) = \frac{1}{3!}\epsilon_{ijk}\epsilon_{lmn}A_{il}A_{jm}A_{kn}$$
$$[\mathbf{A}^{-1}]_{ij} = \frac{1}{2\det(\mathbf{A})}\epsilon_{ipq}\epsilon_{jmn}A_{mp}A_{nq}$$
$$[\mathbf{AB}]_{ik} = A_{ij}B_{jk}$$
$$\mathbf{a} \cdot \mathbf{b} = \delta_{ij}a_i b_j = a_i b_i$$
$$[\mathbf{a} \times \mathbf{b}]_i = \epsilon_{ijk}a_j b_k$$
$$\mathbf{a} \cdot (\mathbf{b} \times \mathbf{c}) = \epsilon_{ijk}a_i b_j c_k$$
$$[\mathbf{a} \times (\mathbf{b} \times \mathbf{c})]_i = \epsilon_{ijk}\epsilon_{klm}a_j b_l c_m$$

   where $\mathbf{A}$ and $\mathbf{B}$ are square matrices and $\mathbf{a}$, $\mathbf{b}$ and $\mathbf{c}$ are vectors.
3. From the tensor definition of $\mathbf{A} \times (\mathbf{B} \times \mathbf{C})$, obtain the tensor definition of $(\mathbf{A} \times \mathbf{B}) \times \mathbf{C}$.
   Answer: We have:

$$[\mathbf{A} \times (\mathbf{B} \times \mathbf{C})]_i = \epsilon_{ijk}\epsilon_{klm}A_j B_l C_m$$
$$[(\mathbf{B} \times \mathbf{C}) \times \mathbf{A}]_i = -\epsilon_{ijk}\epsilon_{klm}A_j B_l C_m$$
$$[(\mathbf{A} \times \mathbf{B}) \times \mathbf{C}]_i = -\epsilon_{ijk}\epsilon_{klm}C_j A_l B_m$$
$$[(\mathbf{A} \times \mathbf{B}) \times \mathbf{C}]_i = \epsilon_{ikj}\epsilon_{klm}C_j A_l B_m$$
$$[(\mathbf{A} \times \mathbf{B}) \times \mathbf{C}]_i = \epsilon_{ikm}\epsilon_{kjl}A_j B_l C_m$$
$$[(\mathbf{A} \times \mathbf{B}) \times \mathbf{C}]_i = \epsilon_{ikm}\epsilon_{jlk}A_j B_l C_m$$

   where line 2 is justified by the anti-commutative property of cross product of vectors [i.e. vector $\mathbf{A}$ and vector $(\mathbf{B} \times \mathbf{C})$], in line 3 we relabel the three vectors, in line 4 we use the anti-symmetric property of the permutation tensor, in line 5 we relabel the dummy indices, and in line 6 we use the cyclic property of the indices of the permutation tensor.

4. We have the following tensors in orthonormal Cartesian coordinates of a 3D space:

$$\mathbf{A} = (22, 3\pi, 6.3) \qquad \mathbf{B} = (3e, 1.8, 4.9) \qquad \mathbf{C} = (47, 5e, 3.5)$$

$$\mathbf{D} = \begin{bmatrix} \pi & 3 \\ 4 & e \end{bmatrix} \qquad \mathbf{E} = \begin{bmatrix} 3 & e^2 \\ \pi^3 & 7 \end{bmatrix}$$

Use the tensor expressions for the relevant mathematical concepts with systematic substitution of the indices values to find the following: $\operatorname{tr}(\mathbf{D})$, $\det(\mathbf{E})$, $\mathbf{D}^{-1}$, $\mathbf{E} \cdot \mathbf{D}$, $\mathbf{A} \cdot \mathbf{C}$, $\mathbf{C} \times \mathbf{B}$, $\mathbf{C} \cdot (\mathbf{A} \times \mathbf{B})$ and $\mathbf{B} \times (\mathbf{C} \times \mathbf{A})$.

Answer:

- $\operatorname{tr}(\mathbf{D})$:
$$\operatorname{tr}(\mathbf{D}) = D_{ii} = D_{11} + D_{22} = \pi + e$$

- $\det(\mathbf{E})$:
$$\begin{aligned} \det(\mathbf{E}) &= \epsilon_{ij} E_{i1} E_{j2} \\ &= \epsilon_{11} E_{11} E_{12} + \epsilon_{22} E_{21} E_{22} + \epsilon_{12} E_{11} E_{22} + \epsilon_{21} E_{21} E_{12} \\ &= 0 + 0 + E_{11} E_{22} - E_{21} E_{12} \\ &= (3 \times 7) - (\pi^3 e^2) \\ &= 21 - \pi^3 e^2 \end{aligned}$$

- $\mathbf{D}^{-1}$: we use the formula:
$$\begin{aligned} \left[\mathbf{D}^{-1}\right]_{ij} &= \frac{1}{\det(\mathbf{D})} \delta_{jn}^{im} D_{nm} \\ &= \frac{1}{\det(\mathbf{D})} \left(\delta_{j1}^{i1} D_{11} + \delta_{j2}^{i1} D_{21} + \delta_{j1}^{i2} D_{12} + \delta_{j2}^{i2} D_{22} \right) \end{aligned}$$

Therefore, we have:

$$\begin{aligned} \left[\mathbf{D}^{-1}\right]_{11} &= \frac{1}{\det(\mathbf{D})} \left(\delta_{11}^{11} D_{11} + \delta_{12}^{11} D_{21} + \delta_{11}^{12} D_{12} + \delta_{12}^{12} D_{22} \right) \\ &= \frac{1}{\det(\mathbf{D})} (0 + 0 + 0 + D_{22}) \\ &= \frac{D_{22}}{\det(\mathbf{D})} \\ \left[\mathbf{D}^{-1}\right]_{12} &= \frac{1}{\det(\mathbf{D})} \left(\delta_{21}^{11} D_{11} + \delta_{22}^{11} D_{21} + \delta_{21}^{12} D_{12} + \delta_{22}^{12} D_{22} \right) \\ &= \frac{1}{\det(\mathbf{D})} (0 + 0 - D_{12} + 0) \\ &= \frac{-D_{12}}{\det(\mathbf{D})} \\ \left[\mathbf{D}^{-1}\right]_{21} &= \frac{1}{\det(\mathbf{D})} \left(\delta_{11}^{21} D_{11} + \delta_{12}^{21} D_{21} + \delta_{11}^{22} D_{12} + \delta_{12}^{22} D_{22} \right) \end{aligned}$$

$$= \frac{1}{\det(\mathbf{D})}(0 - D_{21} + 0 + 0)$$

$$= \frac{-D_{21}}{\det(\mathbf{D})}$$

$$[\mathbf{D}^{-1}]_{22} = \frac{1}{\det(\mathbf{D})}\left(\delta_{21}^{21}D_{11} + \delta_{22}^{21}D_{21} + \delta_{21}^{22}D_{12} + \delta_{22}^{22}D_{22}\right)$$

$$= \frac{1}{\det(\mathbf{D})}(D_{11} + 0 + 0 + 0)$$

$$= \frac{D_{11}}{\det(\mathbf{D})}$$

Now, $\det(\mathbf{D}) = \pi e - 12$ and hence:

$$\mathbf{D}^{-1} = \frac{1}{\det(\mathbf{D})}\begin{bmatrix} D_{22} & -D_{12} \\ -D_{21} & D_{11} \end{bmatrix}$$

$$= \frac{1}{\pi e - 12}\begin{bmatrix} e & -3 \\ -4 & \pi \end{bmatrix}$$

- $\mathbf{E} \cdot \mathbf{D}$: we use the formula:

$$[\mathbf{E} \cdot \mathbf{D}]_{ij} = E_{ik}D_{kj} = E_{i1}D_{1j} + E_{i2}D_{2j}$$

that is:

$$\begin{aligned}[][\mathbf{E} \cdot \mathbf{D}]_{11} &= E_{11}D_{11} + E_{12}D_{21} \\ &= 3\pi + e^2 4 \\ &= 3\pi + 4e^2 \\ [\mathbf{E} \cdot \mathbf{D}]_{12} &= E_{11}D_{12} + E_{12}D_{22} \\ &= 3 \times 3 + e^2 e \\ &= 9 + e^3 \\ [\mathbf{E} \cdot \mathbf{D}]_{21} &= E_{21}D_{11} + E_{22}D_{21} \\ &= \pi^3 \pi + 7 \times 4 \\ &= \pi^4 + 28 \\ [\mathbf{E} \cdot \mathbf{D}]_{22} &= E_{21}D_{12} + E_{22}D_{22} \\ &= \pi^3 3 + 7e \\ &= 3\pi^3 + 7e \end{aligned}$$

Hence:

$$\mathbf{E} \cdot \mathbf{D} = \begin{bmatrix} 3\pi + 4e^2 & 9 + e^3 \\ \pi^4 + 28 & 3\pi^3 + 7e \end{bmatrix}$$

- $\mathbf{A} \cdot \mathbf{C}$:

$$\mathbf{A} \cdot \mathbf{C} = A_i C_i$$

7 TENSORS IN APPLICATION

$$
\begin{aligned}
&= A_1 C_1 + A_2 C_2 + A_3 C_3 \\
&= 22 \times 47 + 3\pi \times 5e + 6.3 \times 3.5 \\
&= 1034 + 15\pi e + 22.05 \\
&\simeq 1184.15
\end{aligned}
$$

- $\mathbf{C} \times \mathbf{B}$: we use the formula:

$$[\mathbf{C} \times \mathbf{B}]_i = \epsilon_{ijk} C_j B_k$$

Now, since ϵ_{ijk} is zero when we have repetitive indices then:
when $i = 1$ we have $\epsilon_{111} = \epsilon_{112} = \epsilon_{113} = \epsilon_{121} = \epsilon_{122} = \epsilon_{131} = \epsilon_{133} = 0$.
when $i = 2$ we have $\epsilon_{211} = \epsilon_{212} = \epsilon_{221} = \epsilon_{222} = \epsilon_{223} = \epsilon_{232} = \epsilon_{233} = 0$.
when $i = 3$ we have $\epsilon_{311} = \epsilon_{313} = \epsilon_{322} = \epsilon_{323} = \epsilon_{331} = \epsilon_{332} = \epsilon_{333} = 0$.
Hence, we should have:

$$
\begin{aligned}
[\mathbf{C} \times \mathbf{B}]_1 &= \epsilon_{123} C_2 B_3 + \epsilon_{132} C_3 B_2 + 0 + 0 + 0 + 0 + 0 + 0 + 0 \\
&= C_2 B_3 - C_3 B_2 \\
&= 5e \times 4.9 - 3.5 \times 1.8 \\
&= 24.5e - 6.3 \\
[\mathbf{C} \times \mathbf{B}]_2 &= \epsilon_{213} C_1 B_3 + \epsilon_{231} C_3 B_1 + 0 + 0 + 0 + 0 + 0 + 0 + 0 \\
&= -C_1 B_3 + C_3 B_1 \\
&= -47 \times 4.9 + 3.5 \times 3e \\
&= -230.3 + 10.5e \\
[\mathbf{C} \times \mathbf{B}]_3 &= \epsilon_{312} C_1 B_2 + \epsilon_{321} C_2 B_1 + 0 + 0 + 0 + 0 + 0 + 0 + 0 \\
&= C_1 B_2 - C_2 B_1 \\
&= 47 \times 1.8 - 5e \times 3e \\
&= 84.6 - 15e^2
\end{aligned}
$$

Hence:

$$
\begin{aligned}
\mathbf{C} \times \mathbf{B} &= \left(24.5e - 6.3,\ 10.5e - 230.3,\ 84.6 - 15e^2\right) \\
&\simeq (60.30, -201.76, -26.24)
\end{aligned}
$$

- $\mathbf{C} \cdot (\mathbf{A} \times \mathbf{B})$: we use the formula:

$$\mathbf{C} \cdot (\mathbf{A} \times \mathbf{B}) = \epsilon_{ijk} C_i A_j B_k$$

Now, since ϵ_{ijk} is zero when we have repetitive indices then we should have only 6 non-vanishing terms which represent the non-repetitive permutations of 123, that is:

$$
\begin{aligned}
\mathbf{C} \cdot (\mathbf{A} \times \mathbf{B}) = {}& \epsilon_{123} C_1 A_2 B_3 + \epsilon_{312} C_3 A_1 B_2 + \epsilon_{231} C_2 A_3 B_1 + \\
& \epsilon_{132} C_1 A_3 B_2 + \epsilon_{213} C_2 A_1 B_3 + \epsilon_{321} C_3 A_2 B_1
\end{aligned}
$$

$$
\begin{aligned}
&= +C_1A_2B_3 + C_3A_1B_2 + C_2A_3B_1 \\
&\quad -C_1A_3B_2 - C_2A_1B_3 - C_3A_2B_1 \\
&= +47 \times 3\pi \times 4.9 + 3.5 \times 22 \times 1.8 + 5e \times 6.3 \times 3e \\
&\quad -47 \times 6.3 \times 1.8 - 5e \times 22 \times 4.9 - 3.5 \times 3\pi \times 3e \\
&= 690.9\pi - 394.38 + 94.5e^2 - 539e - 31.5\pi e \\
&\simeq 740.26
\end{aligned}
$$

- $\mathbf{B} \times (\mathbf{C} \times \mathbf{A})$: we use the formula:

$$[\mathbf{B} \times (\mathbf{C} \times \mathbf{A})]_i = \epsilon_{ijk}\epsilon_{klm}B_jC_lA_m$$

Now, since ϵ_{ijk} is zero when we have repetitive indices then we should have only 6 non-vanishing terms which represent the non-repetitive permutations of 123. This similarly applies to ϵ_{klm}; however the index k of ϵ_{klm}, is fixed by the index k of ϵ_{ijk} and hence we have only 2 non-vanishing permutations for each one of the six permutations of ϵ_{ijk} (i.e. the non-repetitive permutations of lm with $l \neq k$ and $m \neq k$). Accordingly, we should have a total of 12 non-vanishing terms, i.e. 4 non-vanishing terms for each value of i, that is:

$$
\begin{aligned}
[\mathbf{B} \times (\mathbf{C} \times \mathbf{A})]_1 &= \epsilon_{123}\epsilon_{312}B_2C_1A_2 + \epsilon_{132}\epsilon_{213}B_3C_1A_3 + \\
&\quad \epsilon_{123}\epsilon_{321}B_2C_2A_1 + \epsilon_{132}\epsilon_{231}B_3C_3A_1 \\
&= B_2C_1A_2 + B_3C_1A_3 - B_2C_2A_1 - B_3C_3A_1 \\
&= 1.8 \times 47 \times 3\pi + 4.9 \times 47 \times 6.3 - 1.8 \times 5e \times 22 - 4.9 \times 3.5 \times 22 \\
&\simeq 1332.71 \\
[\mathbf{B} \times (\mathbf{C} \times \mathbf{A})]_2 &= \epsilon_{213}\epsilon_{312}B_1C_1A_2 + \epsilon_{231}\epsilon_{123}B_3C_2A_3 + \\
&\quad \epsilon_{213}\epsilon_{321}B_1C_2A_1 + \epsilon_{231}\epsilon_{132}B_3C_3A_2 \\
&= -B_1C_1A_2 + B_3C_2A_3 + B_1C_2A_1 - B_3C_3A_2 \\
&= -3e \times 47 \times 3\pi + 4.9 \times 5e \times 6.3 + 3e \times 5e \times 22 - 4.9 \times 3.5 \times 3\pi \\
&\simeq -915.99 \\
[\mathbf{B} \times (\mathbf{C} \times \mathbf{A})]_3 &= \epsilon_{312}\epsilon_{213}B_1C_1A_3 + \epsilon_{321}\epsilon_{123}B_2C_2A_3 + \\
&\quad \epsilon_{312}\epsilon_{231}B_1C_3A_1 + \epsilon_{321}\epsilon_{132}B_2C_3A_2 \\
&= -B_1C_1A_3 - B_2C_2A_3 + B_1C_3A_1 + B_2C_3A_2 \\
&= -3e \times 47 \times 6.3 - 1.8 \times 5e \times 6.3 + 3e \times 3.5 \times 22 + 1.8 \times 3.5 \times 3\pi \\
&\simeq -1881.48
\end{aligned}
$$

Hence:
$$\mathbf{B} \times (\mathbf{C} \times \mathbf{A}) \simeq (1332.71,\ -915.99,\ -1881.48)$$

5. State the matrix and tensor definitions of the main three independent scalar invariants (I, II and III) of rank-2 tensors.
Answer: They are:

$$I = \mathrm{tr}(\mathbf{A}) = A_{ii}$$

7   TENSORS IN APPLICATION   135

$$
\begin{aligned}
II &= \text{tr}\left(\mathbf{A}^2\right) = A_{ij}A_{ji} \\
III &= \text{tr}\left(\mathbf{A}^3\right) = A_{ij}A_{jk}A_{ki}
\end{aligned}
$$

where $\mathbf{A}$ is a rank-2 tensor.

6. Express the main three independent scalar invariants (I, II and III) of rank-2 tensors in terms of the three subsidiary scalar invariants (I_1, I_2 and I_3).
   Answer: They are:[34]

$$
\begin{aligned}
I &= I_1 \\
II &= I_1^2 - 2I_2 \\
III &= I_1^3 - 3I_1 I_2 + 3I_3
\end{aligned}
$$

7. Referring to question 4, find the three scalar invariants (I, II and III) of $\mathbf{D}$ and the three scalar invariants (I_1, I_2 and I_3) of $\mathbf{E}$ using the tensor definitions of these invariants with systematic index substitution.
   Answer:
   - We have:

$$
\begin{aligned}
I(\mathbf{D}) &= D_{ii} \\
&= D_{11} + D_{22} \\
&= \pi + e \\
&\simeq 5.86 \\
II(\mathbf{D}) &= D_{ij}D_{ji} \\
&= D_{11}D_{11} + D_{12}D_{21} + D_{21}D_{12} + D_{22}D_{22} \\
&= \pi \times \pi + 3 \times 4 + 4 \times 3 + e \times e \\
&= \pi^2 + 24 + e^2 \\
&\simeq 41.26 \\
III(\mathbf{D}) &= D_{ij}D_{jk}D_{ki} \\
&= D_{11}D_{11}D_{11} + D_{12}D_{21}D_{11} + D_{11}D_{12}D_{21} + D_{12}D_{22}D_{21} + \\
&\quad D_{21}D_{11}D_{12} + D_{22}D_{21}D_{12} + D_{21}D_{12}D_{22} + D_{22}D_{22}D_{22} \\
&= \pi^3 + 12\pi + 12\pi + 12e + 12\pi + 12e + 12e + e^3 \\
&= \pi^3 + 36\pi + 36e + e^3 \\
&\simeq 262.05
\end{aligned}
$$

   - We have:

$$
\begin{aligned}
I_1(\mathbf{E}) &= E_{ii} \\
&= E_{11} + E_{22} \\
&= 3 + 7
\end{aligned}
$$

---
[34] We note that some of these definitions may belong to 3D specifically.

$$
\begin{aligned}
&= 10\\
I_2(\mathbf{E}) &= \frac{1}{2}(E_{ii}E_{jj} - E_{ij}E_{ji})\\
&= \frac{1}{2}(E_{11}E_{11} + E_{11}E_{22} + E_{22}E_{11} + E_{22}E_{22}) -\\
&\quad \frac{1}{2}(E_{11}E_{11} + E_{12}E_{21} + E_{21}E_{12} + E_{22}E_{22})\\
&= \frac{1}{2}(9 + 21 + 21 + 49) - \frac{1}{2}(9 + e^2\pi^3 + \pi^3 e^2 + 49)\\
&= 21 - e^2\pi^3\\
&\simeq -208.11\\
I_3(\mathbf{E}) &= \det(\mathbf{E})\\
&= \epsilon_{ij}E_{i1}E_{j2}\\
&= \epsilon_{11}E_{11}E_{12} + \epsilon_{22}E_{21}E_{22} + \epsilon_{12}E_{11}E_{22} + \epsilon_{21}E_{21}E_{12}\\
&= 0 + 0 + E_{11}E_{22} - E_{21}E_{12}\\
&= (3 \times 7) - (\pi^3 e^2)\\
&= 21 - \pi^3 e^2\\
&\simeq -208.11
\end{aligned}
$$

8. State the following vector identities in tensor notation:

$$
\begin{aligned}
\nabla \times \mathbf{r} &= \mathbf{0}\\
\nabla \cdot (f\mathbf{A}) &= f\nabla \cdot \mathbf{A} + \mathbf{A} \cdot \nabla f\\
\mathbf{A} \times (\nabla \times \mathbf{B}) &= (\nabla \mathbf{B}) \cdot \mathbf{A} - \mathbf{A} \cdot \nabla \mathbf{B}\\
\nabla \times (\mathbf{A} \times \mathbf{B}) &= (\mathbf{B} \cdot \nabla)\mathbf{A} + (\nabla \cdot \mathbf{B})\mathbf{A} - (\nabla \cdot \mathbf{A})\mathbf{B} - (\mathbf{A} \cdot \nabla)\mathbf{B}
\end{aligned}
$$

Answer: Assuming Cartesian coordinates, we have:

$$
\begin{aligned}
\epsilon_{ijk}\partial_j x_k &= 0\\
\partial_i(fA_i) &= f\partial_i A_i + A_i \partial_i f\\
\epsilon_{ijk}\epsilon_{klm}A_j \partial_l B_m &= (\partial_i B_m)A_m - A_l(\partial_l B_i)\\
\epsilon_{ijk}\epsilon_{klm}\partial_j(A_l B_m) &= (B_m \partial_m)A_i + (\partial_m B_m)A_i - (\partial_j A_j)B_i - (A_j \partial_j)B_i
\end{aligned}
$$

9. State the divergence and Stokes theorems for a vector field in Cartesian coordinates using vector and tensor notations. Also, define all the symbols involved.
   Answer: Assuming Cartesian coordinates, we have:
   • Divergence theorem:

$$
\begin{aligned}
\iiint_\Omega \nabla \cdot \mathbf{A}\, d\tau &= \iint_S \mathbf{A} \cdot \mathbf{n}\, d\sigma\\
\int_\Omega \partial_i A_i\, d\tau &= \int_S A_i n_i\, d\sigma
\end{aligned}
$$

where **A** is a differentiable vector field, Ω is a bounded region in an nD space enclosed by a generalized surface S, $d\tau$ and $d\sigma$ are generalized volume and area differentials, **n** and n_i are the unit vector normal to the surface and its ith component, and the index i ranges over $1, \ldots, n$.
• Stokes theorem:

$$\iint_S (\nabla \times \mathbf{A}) \cdot \mathbf{n} \, d\sigma = \int_C \mathbf{A} \cdot d\mathbf{r}$$

$$\int_S \epsilon_{ijk} \partial_j A_k n_i d\sigma = \int_C A_i dx_i$$

where C stands for the perimeter of the surface S, and $d\mathbf{r}$ is a differential of the position vector which is tangent to the perimeter, x_i is a Cartesian coordinate while the other symbols are as defined in the first part.

10. Prove the following vector identities using tensor notation and techniques with full justification of each step:

$$\nabla \cdot \mathbf{r} = n$$
$$\nabla \cdot (\nabla \times \mathbf{A}) = 0$$
$$\mathbf{A} \cdot (\mathbf{B} \times \mathbf{C}) = \mathbf{C} \cdot (\mathbf{A} \times \mathbf{B})$$
$$\nabla \times (\nabla \times \mathbf{A}) = \nabla (\nabla \cdot \mathbf{A}) - \nabla^2 \mathbf{A}$$

Answer: Assuming Cartesian coordinates, we have:
• $\nabla \cdot \mathbf{r} = n$:

$$\begin{aligned} \nabla \cdot \mathbf{r} &= \partial_i x_i \\ &= \delta_{ii} \\ &= n \end{aligned}$$

where line 1 is the definition of divergence, line 2 is the identity $\partial_j x_i = \delta_{ij}$ with $j = i$, and line 3 is the identity $\delta_{ii} = n$ which is given in the book and proved in Exercise 22 of § 4.
• $\nabla \cdot (\nabla \times \mathbf{A}) = 0$:

$$\begin{aligned} \nabla \cdot (\nabla \times \mathbf{A}) &= \partial_i \left[\nabla \times \mathbf{A}\right]_i \\ &= \partial_i \left(\epsilon_{ijk} \partial_j A_k\right) \\ &= \epsilon_{ijk} \partial_i \partial_j A_k \\ &= \epsilon_{ijk} \partial_j \partial_i A_k \\ &= -\epsilon_{jik} \partial_j \partial_i A_k \\ &= -\epsilon_{ijk} \partial_i \partial_j A_k \\ &= 0 \end{aligned}$$

where line 1 is the definition of divergence, line 2 is the definition of curl, line 3 is the constancy of components of the permutation tensor, line 4 is the commutativity

of partial differential operators, line 5 is the anti-symmetry of the permutation tensor, line 6 is relabeling of dummy indices, and line 7 is based on comparing line 6 with line 3 plus the fact that only 0 is equal to its negative.
- $\mathbf{A} \cdot (\mathbf{B} \times \mathbf{C}) = \mathbf{C} \cdot (\mathbf{A} \times \mathbf{B})$:

$$
\begin{aligned}
\mathbf{A} \cdot (\mathbf{B} \times \mathbf{C}) &= \epsilon_{ijk} A_i B_j C_k \\
&= \epsilon_{kij} A_i B_j C_k \\
&= \epsilon_{kij} C_k A_i B_j \\
&= \mathbf{C} \cdot (\mathbf{A} \times \mathbf{B})
\end{aligned}
$$

where line 1 is the definition of scalar triple product, line 2 is the cyclic property of ϵ_{ijk}, line 3 is the commutativity of ordinary multiplication, and line 4 is the definition of scalar triple product.
- $\nabla \times (\nabla \times \mathbf{A}) = \nabla (\nabla \cdot \mathbf{A}) - \nabla^2 \mathbf{A}$:

$$
\begin{aligned}
[\nabla \times (\nabla \times \mathbf{A})]_i &= \epsilon_{ijk} \partial_j [\nabla \times \mathbf{A}]_k \\
&= \epsilon_{ijk} \partial_j (\epsilon_{klm} \partial_l A_m) \\
&= \epsilon_{ijk} \epsilon_{klm} \partial_j (\partial_l A_m) \\
&= \epsilon_{ijk} \epsilon_{lmk} \partial_j \partial_l A_m \\
&= (\delta_{il} \delta_{jm} - \delta_{im} \delta_{jl}) \partial_j \partial_l A_m \\
&= \delta_{il} \delta_{jm} \partial_j \partial_l A_m - \delta_{im} \delta_{jl} \partial_j \partial_l A_m \\
&= \partial_m \partial_i A_m - \partial_l \partial_l A_i \\
&= \partial_i (\partial_m A_m) - \partial_{ll} A_i \\
&= [\nabla (\nabla \cdot \mathbf{A})]_i - [\nabla^2 \mathbf{A}]_i \\
&= [\nabla (\nabla \cdot \mathbf{A}) - \nabla^2 \mathbf{A}]_i
\end{aligned}
$$

where line 1 is the definition of curl (first curl), line 2 is the definition of curl (second curl), line 3 is the constancy of components of the permutation tensor, line 4 is the cyclic property of ϵ_{klm}, line 5 is the epsilon-delta identity, line 6 is the distributivity of product over algebraic sum, line 7 is index replacement operation, line 8 is the commutativity of partial differential operators and the definition of second derivative, line 9 is based on the definitions of gradient, divergence and Laplacian, line 10 is the distributivity of indexing over algebraic sum of tensor terms. Since i is a free index, then the identity equally applies to all components and hence:

$$\nabla \times (\nabla \times \mathbf{A}) = \nabla (\nabla \cdot \mathbf{A}) - \nabla^2 \mathbf{A}$$

as required.

11. What is the type, in the form of (m, n, w), of the Riemann-Christoffel curvature tensor of the first and second kinds?
    Answer: The type of the first kind is $(0, 4, 0)$ while the type of the second kind is $(1, 3, 0)$.

7 TENSORS IN APPLICATION   139

12. What are the other names used to label the Riemann-Christoffel curvature tensor of the first and second kinds?
    Answer: The first kind may be called the covariant or totally covariant Riemann-Christoffel curvature tensor, while the second kind may be called the mixed Riemann-Christoffel curvature tensor.
13. What is the importance of the Riemann-Christoffel curvature tensor with regard to characterizing the space as flat or curved?
    Answer: The Riemann-Christoffel curvature tensor vanishes identically *iff* the space is globally flat. Hence, by testing the Riemann-Christoffel curvature tensor we can determine if the space is flat (if the tensor vanishes identically) or curved (if not).
14. State the mathematical definition of the Riemann-Christoffel curvature tensor of either kinds in determinantal form.
    Answer:

$$R_{ijkl} = \begin{vmatrix} \partial_k & \partial_l \\ [jk,i] & [jl,i] \end{vmatrix} + \begin{vmatrix} \Gamma^r_{jk} & \Gamma^r_{jl} \\ [ik,r] & [il,r] \end{vmatrix}$$

$$R^i_{jkl} = \begin{vmatrix} \partial_k & \partial_l \\ \Gamma^i_{jk} & \Gamma^i_{jl} \end{vmatrix} + \begin{vmatrix} \Gamma^r_{jl} & \Gamma^r_{jk} \\ \Gamma^i_{rl} & \Gamma^i_{rk} \end{vmatrix}$$

15. How can we obtain the Riemann-Christoffel curvature tensor of the first kind from the second kind and vice versa?
    Answer: The first kind can be obtained from the second kind by lowering the first index of the second kind, while the second kind can be obtained from the first kind by raising the first index of the first kind.
16. Using the definition of the second order mixed covariant derivative of a vector field and the definition of the mixed Riemann-Christoffel curvature tensor, verify the following equation: $A_{j;kl} - A_{j;lk} = R^i_{\ jkl} A_i$. Repeat the question with the equation: $A^j_{\ ;kl} - A^j_{\ ;lk} = R^j_{\ ilk} A^i$.
    Answer:
    - In question 52 of § 5 we obtained:

$$\begin{aligned} A_{i;jk} - A_{i;kj} &= -A_b \partial_k \Gamma^b_{ij} + \Gamma^a_{ik} \Gamma^b_{aj} A_b + A_b \partial_j \Gamma^b_{ik} - \Gamma^a_{ij} \Gamma^b_{ak} A_b \\ &= \left(\partial_j \Gamma^b_{ik} - \partial_k \Gamma^b_{ij} + \Gamma^a_{ik} \Gamma^b_{aj} - \Gamma^a_{ij} \Gamma^b_{ak} \right) A_b \\ &= R^b_{\ ijk} A_b \end{aligned}$$

On relabeling the indices, we obtain:

$$A_{j;kl} - A_{j;lk} = R^i_{\ jkl} A_i$$

which is the required result.
    - In the book we obtained the following equations:

$$A^i_{\ ;jk} = \partial_k \partial_j A^i + \Gamma^i_{aj} \partial_k A^a - \Gamma^a_{jk} \partial_a A^i + \Gamma^i_{ak} \partial_j A^a + A^a \left(\partial_k \Gamma^i_{aj} - \Gamma^b_{jk} \Gamma^i_{ba} + \Gamma^i_{bk} \Gamma^b_{aj} \right)$$

$$A^i_{;kj} \;=\; \partial_j\partial_k A^i + \Gamma^i_{ak}\partial_j A^a - \Gamma^a_{kj}\partial_a A^i + \Gamma^i_{aj}\partial_k A^a + A^a\left(\partial_j\Gamma^i_{ak} - \Gamma^b_{kj}\Gamma^i_{ba} + \Gamma^i_{bj}\Gamma^b_{ak}\right)$$

On taking the difference we obtain:

$$\begin{aligned}
A^i_{;jk} - A^i_{;kj} \\
&= \partial_k\partial_j A^i + \Gamma^i_{aj}\partial_k A^a - \Gamma^a_{jk}\partial_a A^i + \Gamma^i_{ak}\partial_j A^a + A^a\left(\partial_k\Gamma^i_{aj} - \Gamma^b_{jk}\Gamma^i_{ba} + \Gamma^i_{bk}\Gamma^b_{aj}\right) - \\
&\quad \left[\partial_j\partial_k A^i + \Gamma^i_{ak}\partial_j A^a - \Gamma^a_{kj}\partial_a A^i + \Gamma^i_{aj}\partial_k A^a + A^a\left(\partial_j\Gamma^i_{ak} - \Gamma^b_{kj}\Gamma^i_{ba} + \Gamma^i_{bj}\Gamma^b_{ak}\right)\right] \\
&= A^a\left(\partial_k\Gamma^i_{aj} + \Gamma^i_{bk}\Gamma^b_{aj}\right) - A^a\left(\partial_j\Gamma^i_{ak} + \Gamma^i_{bj}\Gamma^b_{ak}\right) \\
&= \left(\partial_k\Gamma^i_{aj} - \partial_j\Gamma^i_{ak} + \Gamma^i_{bk}\Gamma^b_{aj} - \Gamma^i_{bj}\Gamma^b_{ak}\right)A^a \\
&= R^i_{akj}A^a
\end{aligned}$$

On relabeling the indices, we obtain:

$$A^j_{;kl} - A^j_{;lk} = R^j_{ilk}A^i$$

which is the required result.

17. Based on the equations in question 16, what is the necessary and sufficient condition for the covariant differential operators to become commutative?
    Answer: The covariant differential operators become commutative *iff* the Riemann-Christoffel curvature tensor vanishes identically.

18. State, mathematically, the anti-symmetric and block symmetric properties of the Riemann-Christoffel curvature tensor of the first kind in its four indices.
    Answer:

$$\begin{aligned}
R_{jikl} &= -R_{ijkl} \\
R_{ijlk} &= -R_{ijkl} \\
R_{klij} &= +R_{ijkl}
\end{aligned}$$

where line 1 is the anti-symmetric property in the first two indices, line 2 is the anti-symmetric property in the last two indices, and line 3 is the block symmetric property.

19. Based on the two anti-symmetric properties of the covariant Riemann-Christoffel curvature tensor, list all the forms of the components of the tensor that are identically zero (e.g. R_{iijk}).
    Answer: Any component with at least a pair of identical anti-symmetric indices should vanish identically. Hence, the forms of the components that are identically zero are:

    Two indices identical: R_{iijk} R_{ijkk} R_{iikk}

    Three indices identical: R_{iiik} R_{iiji} R_{kjkk} R_{ikkk}

    Four indices identical: R_{iiii}

20. Verify the block symmetric property and the two anti-symmetric properties of the covariant Riemann-Christoffel curvature tensor using its definition.
    Answer:

- Block symmetric property:

$$
\begin{aligned}
R_{klij} &= \frac{1}{2}\left(\partial_i\partial_l g_{jk} + \partial_j\partial_k g_{li} - \partial_i\partial_k g_{lj} - \partial_j\partial_l g_{ik}\right) + g^{rs}\left([kj,r][li,s] - [ki,r][lj,s]\right) \\
&= \frac{1}{2}\left(\partial_l\partial_i g_{jk} + \partial_k\partial_j g_{li} - \partial_k\partial_i g_{lj} - \partial_l\partial_j g_{ik}\right) + g^{rs}\left([kj,r][li,s] - [ki,r][lj,s]\right) \\
&= \frac{1}{2}\left(\partial_k\partial_j g_{li} + \partial_l\partial_i g_{jk} - \partial_k\partial_i g_{lj} - \partial_l\partial_j g_{ik}\right) + g^{rs}\left([li,s][kj,r] - [ki,r][lj,s]\right) \\
&= \frac{1}{2}\left(\partial_k\partial_j g_{li} + \partial_l\partial_i g_{jk} - \partial_k\partial_i g_{jl} - \partial_l\partial_j g_{ki}\right) + g^{rs}\left([il,s][jk,r] - [ik,r][jl,s]\right) \\
&= \frac{1}{2}\left(\partial_k\partial_j g_{li} + \partial_l\partial_i g_{jk} - \partial_k\partial_i g_{jl} - \partial_l\partial_j g_{ki}\right) + g^{rs}\left([il,r][jk,s] - [ik,r][jl,s]\right) \\
&= R_{ijkl}
\end{aligned}
$$

where line 1 is obtained from the given definition of the covariant Riemann-Christoffel curvature tensor (which is given in the book as R_{ijkl}) with relabeling the indices ($ijkl \to klij$), line 2 is the commutativity of partial differential operators, line 3 is reordering of terms and factors, line 4 is the symmetry of the metric tensor and the symmetry of the Christoffel symbols in their paired indices, line 5 is relabeling of dummy indices plus the symmetry of the metric tensor, and line 6 is the definition of R_{ijkl} which is given in the book.

- Anti-symmetric property in the first two indices:

$$
\begin{aligned}
R_{jikl} &= \frac{1}{2}\left(\partial_k\partial_i g_{lj} + \partial_l\partial_j g_{ik} - \partial_k\partial_j g_{il} - \partial_l\partial_i g_{kj}\right) + g^{rs}\left([jl,r][ik,s] - [jk,r][il,s]\right) \\
&= -\left[\frac{1}{2}\left(\partial_k\partial_j g_{il} + \partial_l\partial_i g_{kj} - \partial_k\partial_i g_{lj} - \partial_l\partial_j g_{ik}\right) + g^{rs}\left([jk,r][il,s] - [jl,r][ik,s]\right)\right] \\
&= -\left[\frac{1}{2}\left(\partial_k\partial_j g_{li} + \partial_l\partial_i g_{jk} - \partial_k\partial_i g_{jl} - \partial_l\partial_j g_{ki}\right) + g^{sr}\left([il,s][jk,r] - [ik,s][jl,r]\right)\right] \\
&= -\left[\frac{1}{2}\left(\partial_k\partial_j g_{li} + \partial_l\partial_i g_{jk} - \partial_k\partial_i g_{jl} - \partial_l\partial_j g_{ki}\right) + g^{rs}\left([il,r][jk,s] - [ik,r][jl,s]\right)\right] \\
&= -R_{ijkl}
\end{aligned}
$$

where line 1 is obtained from the given definition of the covariant Riemann-Christoffel curvature tensor with relabeling the indices ($ijkl \to jikl$), line 2 is taking a common factor of -1, line 3 is the symmetry of the metric tensor plus reordering, line 4 is relabeling of dummy indices, and line 5 is the definition of R_{ijkl} which is given in the book.

- Anti-symmetric property in the last two indices:

$$
\begin{aligned}
R_{ijlk} &= \frac{1}{2}\left(\partial_l\partial_j g_{ki} + \partial_k\partial_i g_{jl} - \partial_l\partial_i g_{jk} - \partial_k\partial_j g_{li}\right) + g^{rs}\left([ik,r][jl,s] - [il,r][jk,s]\right) \\
&= -\left[\frac{1}{2}\left(\partial_l\partial_i g_{jk} + \partial_k\partial_j g_{li} - \partial_l\partial_j g_{ki} - \partial_k\partial_i g_{jl}\right) + g^{rs}\left([il,r][jk,s] - [ik,r][jl,s]\right)\right]
\end{aligned}
$$

$$= -\left[\frac{1}{2}\left(\partial_k\partial_j g_{li} + \partial_l\partial_i g_{jk} - \partial_k\partial_i g_{jl} - \partial_l\partial_j g_{ki}\right) + g^{rs}\left([il,r][jk,s] - [ik,r][jl,s]\right)\right]$$

$$= -R_{ijkl}$$

where line 1 is obtained from the given definition of the covariant Riemann-Christoffel curvature tensor with relabeling the indices ($ijkl \to ijlk$), line 2 is taking a common factor of -1, line 3 is reordering of terms, and line 4 is the definition of R_{ijkl} which is given in the book.

21. Repeat question 20 for the anti-symmetric property of the mixed Riemann-Christoffel curvature tensor in its last two indices.

    Answer:

    $$\begin{aligned}R^i_{jlk} &= \partial_l \Gamma^i_{jk} - \partial_k \Gamma^i_{jl} + \Gamma^r_{jk}\Gamma^i_{rl} - \Gamma^r_{jl}\Gamma^i_{rk} \\ &= -\left(\partial_k \Gamma^i_{jl} - \partial_l \Gamma^i_{jk} + \Gamma^r_{jl}\Gamma^i_{rk} - \Gamma^r_{jk}\Gamma^i_{rl}\right) \\ &= -R^i_{jkl}\end{aligned}$$

    where line 1 is obtained from the given definition of the mixed Riemann-Christoffel curvature tensor (which is given in the book as R^i_{jkl}) with relabeling the indices ($ijkl \to ijlk$), line 2 is taking a common factor of -1, and line 3 is the definition of R^i_{jkl} which is given in the book.

22. Based on the block symmetric and anti-symmetric properties of the covariant Riemann-Christoffel curvature tensor, find (with full justification) the number of distinct non-vanishing entries of the three main types of this tensor (see the equations for N_2, N_3 and N_4 which are given in the book). Hence, find the total number of the independent non-zero components of this tensor.

    Answer: We have three main cases:

    (a) Entries with only two distinct indices of type R_{ijij}: due to the anti-symmetric properties, the indices 1 and 2 should be different and the indices 3 and 4 should be different so that the component does not vanish identically. Moreover, since we have only two distinct indices then the form of this type should be either R_{ijij} or R_{ijji}. However, since these two forms differ only by sign due to the anti-symmetric property in the last two indices (or the first two indices), then all the independent non-zero components of this type can be represented by just one of these forms, say R_{ijij}. Now, the number of components of this form is equal to the number of permutations of ij and because i and j are distinct then the number of permutations is $n(n-1)$. However, due to the anti-symmetric properties the permutations corresponding to IJ (say 23) and the permutations corresponding to JI (say 32) are identical and hence they are not distinct. Therefore, only half of these permutations will contribute to the number of independent non-zero components of this tensor, that is:[35]

    $$N_2 = \frac{n(n-1)}{2}$$

---

[35] We note that the given formula applies even to $n = 1$ since it gives $N_2 = 0$.

7 TENSORS IN APPLICATION

(b) Entries with only three distinct indices of type R_{ijki}: due to the anti-symmetric properties, the indices 1 and 2 should be different and the indices 3 and 4 should be different so that the component does not vanish identically. Therefore, the identical indices should be either 1 and 3 (i.e. R_{ijik}), or 1 and 4 (i.e. R_{ijki}), or 2 and 3 (i.e. R_{jiik}), or 2 and 4 (i.e. R_{jiki}). Now, due to anti-symmetry the forms R_{ijik} and R_{ijki} are not independent and the forms R_{jiik} and R_{jiki} are not independent and hence we are left with R_{ijki} (representing the first two forms) and R_{jiik} (representing the last two forms). However, due to anti-symmetry we have $R_{ijki} = -R_{jiki} = R_{jiik}$ and hence even these two forms are not independent. Therefore, all the independent non-zero components with only three distinct indices are represented by a single form, say R_{ijki}. Now, since the last i is not independent (and hence all the components represented by R_{ijki} can be similarly represented by R_{ijk}) then the number of components represented by this form is equal to the number of permutations of ijk and since these three indices are distinct then we should have $n(n-1)(n-2)$ permutations. However, due to the anti-symmetric property in the first two indices then $R_{ijk} = -R_{jik}$ (e.g. $R_{123} = -R_{213}$) and hence only half of these permutations will contribute to the number of independent non-zero components of this tensor, that is:[36]

$$N_3 = \frac{n(n-1)(n-2)}{2}$$

(c) Entries with four distinct indices of type R_{ijkl}: the number of components represented by R_{ijkl} is equal to the number of permutations of $ijkl$ which is given by $n(n-1)(n-2)(n-3)$ since all these indices are distinct. However, due to the anti-symmetric property in the first two indices we should halve this number since the shift in indices produces only difference in sign and hence the components that are opposite in sign are not independent. This similarly applies to the anti-symmetric property in the last two indices and hence we should halve again. So, we are left with $[n(n-1)(n-2)(n-3)]/4$ non-vanishing and potentially independent components. Now, the block symmetric property will reduce this number by a factor of 2 because all the remaining components have no anti-symmetric correspondence since it is already removed and hence the two blocks are like two indices and hence by the block symmetry we should also halve. This means that we are left with $[n(n-1)(n-2)(n-3)]/8$ non-vanishing and potentially independent components. Finally, the first Bianchi identity links each set of three of the remaining components and hence one of the three can be expressed in terms of the other two and therefore it is not independent. This means that the number will be reduced by a factor of $2/3$ (since one third of the three is dependent on the two thirds). Hence, the number of the independent non-zero components that are contributed by this type is:[37]

$$N_4 = \frac{n(n-1)(n-2)(n-3)}{8} \times \frac{2}{3} = \frac{n(n-1)(n-2)(n-3)}{12}$$

---

[36] We note that the given formula applies even to $n=1$ and $n=2$ since it gives $N_3 = 0$.

[37] We note that the given formula applies even to $n=1$, $n=2$ and $n=3$ since it gives $N_4 = 0$.

7 *TENSORS IN APPLICATION*   144

To clarify case (c), we present in the following table how N_4 is obtained for the case of 4D space (i.e. $n = 4$) where in column 1 we include all the permutations [i.e. $n(n-1)(n-2)(n-3) = 24$], in column 2 we apply anti-symmetry in the first two indices [i.e. $n(n-1)(n-2)(n-3)/2 = 12$], in column 3 we apply anti-symmetry in the last two indices [i.e. $n(n-1)(n-2)(n-3)/4 = 6$], in column 4 we apply block symmetry [i.e. $n(n-1)(n-2)(n-3)/8 = 3$], and in column 5 we apply the first Bianchi identity [i.e. $n(n-1)(n-2)(n-3)/12 = 2$] since $R_{1324} = R_{1234} + R_{1423}$.[38]

1234	1234	1234	1234	1234
1243	1243			
1324	1324	1324	1324	
1342	1342			
1423	1423	1423	1423	1423
1432	1432			
2134				
2143				
2314	2314	2314		
2341	2341			
2413	2413	2413		
2431	2431			
3124				
3142				
3214				
3241				
3412	3412	3412		
3421	3421			
4123				
4132				
4213				
4231				
4312				
4321				

Accordingly, the total number of the independent non-zero components of this tensor is:

$$N_{\mathrm{RI}} = N_2 + N_3 + N_4 =$$

---

[38] This can be shown as follows:

$$\begin{aligned} R_{1234} + R_{1423} + R_{1342} &= 0 \\ R_{1234} + R_{1423} - R_{1324} &= 0 \\ R_{1324} &= R_{1234} + R_{1423} \end{aligned}$$

where line 1 is the first Bianchi identity and line 2 is anti-symmetry in the last two indices.

$$= \frac{n(n-1)}{2} + \frac{n(n-1)(n-2)}{2} + \frac{n(n-1)(n-2)(n-3)}{12}$$

$$= \frac{n^2-n}{2} + \frac{n^3-n^2-2n^2+2n}{2} + \frac{(n^3-n^2-2n^2+2n)(n-3)}{12}$$

$$= \frac{n^3-2n^2+n}{2} + \frac{(n^3-3n^2+2n)(n-3)}{12}$$

$$= \frac{n^3-2n^2+n}{2} + \frac{n^4-3n^3+2n^2-3n^3+9n^2-6n}{12}$$

$$= \frac{6n^3-12n^2+6n+n^4-3n^3+2n^2-3n^3+9n^2-6n}{12}$$

$$= \frac{n^4-n^2}{12}$$

$$= \frac{n^2(n^2-1)}{12}$$

23. Use the formulae found in question 22 and other formulae given in the text to find the number of all components, the number of non-zero components, the number of zero components and the number of independent non-zero components of the covariant Riemann-Christoffel curvature tensor in 2D, 3D and 4D spaces.
    Answer: We symbolize the number of all components with N_a, the number of non-zero components with N_{nz}, the number of zero components with N_z and the number of independent non-zero components with N_{RI}, while we use n to symbolize the dimensionality of the space and hence we have:[39]

	$N_a = n^4$	$N_{nz} = n^2(n-1)^2$	$N_z = n^2(2n-1)$	$N_{RI} = \frac{n^2(n^2-1)}{12}$
2D	16	4	12	1
3D	81	36	45	6
4D	256	144	112	20

24. Prove the following identity with full justification of each step of your proof: $R^a_{akl} = 0$.
    Answer: We have:

    $$\begin{aligned} R^a_{akl} &= \partial_k \Gamma^a_{al} - \partial_l \Gamma^a_{ak} + \Gamma^r_{al}\Gamma^a_{rk} - \Gamma^r_{ak}\Gamma^a_{rl} \\ &= \partial_k \Gamma^a_{al} - \partial_l \Gamma^a_{ak} + \Gamma^r_{al}\Gamma^a_{rk} - \Gamma^a_{rk}\Gamma^r_{al} \\ &= \partial_k \Gamma^a_{al} - \partial_l \Gamma^a_{ak} \\ &= \partial_k [\partial_l (\ln \sqrt{g})] - \partial_l [\partial_k (\ln \sqrt{g})] \\ &= \partial_k \partial_l (\ln \sqrt{g}) - \partial_l \partial_k (\ln \sqrt{g}) \\ &= \partial_k \partial_l (\ln \sqrt{g}) - \partial_k \partial_l (\ln \sqrt{g}) \\ &= 0 \end{aligned}$$

    where line 1 is the definition of R^i_{jkl} (which is given in the book) with $i = j = a$, line 2 is relabeling of dummy indices in the last term, line 4 is the identity $\Gamma^j_{ji} = \partial_i (\ln \sqrt{g})$ which is given in the book, and line 6 is the commutativity of partial differential operators.

---

[39] We note that some of the following formulae are not given in the text or in the previous question.

7 TENSORS IN APPLICATION

25. Make a list of all the main properties of the Riemann-Christoffel curvature tensor (i.e. rank, type, symmetry, etc.).
    Answer: Some of these properties are:
    • It is absolute tensor.
    • It is rank-4 tensor.
    • It can be covariant of type $(0,4)$ or mixed of type $(1,3)$.
    • The covariant type is anti-symmetric in its first two indices and in its last two indices and block symmetric in the first and second pairs of indices, while the mixed type is anti-symmetric in its last two indices.
    • It depends only on the metric tensor.
    • It characterizes the space and hence it is used for example as a test for the curvature of space since it vanishes identically *iff* the space is flat.
    • When it vanishes the covariant differential operators become commutative.

26. Prove the following identity using the Bianchi identities: $R_{ijkl;s} + R_{iljk;s} = R_{iksl;j} + R_{ikjs;l}$.
    Answer: We have:
    $$\begin{aligned} R_{ijkl;m} + R_{ijlm;k} + R_{ijmk;l} &= 0 \\ R_{ikjs;l} + R_{iksl;j} + R_{iklj;s} &= 0 \\ -R_{iklj;s} &= R_{iksl;j} + R_{ikjs;l} \\ (-R_{iklj})_{;s} &= R_{iksl;j} + R_{ikjs;l} \\ (R_{ijkl} + R_{iljk})_{;s} &= R_{iksl;j} + R_{ikjs;l} \\ R_{ijkl;s} + R_{iljk;s} &= R_{iksl;j} + R_{ikjs;l} \end{aligned}$$

    where line 1 is the second of the Bianchi identities as given in the book, line 2 is relabeling the indices (i.e. $ijklm \to ikjsl$), line 5 is the first Bianchi identity (i.e. $R_{ijkl} + R_{iljk} + R_{iklj} = 0$), and line 6 is the distributivity of covariant derivative.

27. Write the first Bianchi identity in its first and second kinds.
    Answer: The first and second kinds of the first Bianchi identity are given respectively by:
    $$\begin{aligned} R_{ijkl} + R_{iljk} + R_{iklj} &= 0 \\ R^i{}_{jkl} + R^i{}_{ljk} + R^i{}_{klj} &= 0 \end{aligned}$$

    where the indexed R are the covariant and mixed type Riemann-Christoffel curvature tensor.

28. Verify the following form of the first Bianchi identity using the mathematical definition of the Riemann-Christoffel curvature tensor: $R_{ijkl} + R_{kijl} + R_{jkil} = 0$.
    Answer: We have:
    $$\begin{aligned} R_{ijkl} + R_{kijl} + R_{jkil} &= R_{ijkl} + R_{kijl} + R_{iljk} \\ &= R_{ijkl} - R_{ikjl} + R_{iljk} \\ &= R_{ijkl} + R_{iklj} + R_{iljk} \\ &= \partial_k [jl,i] - \partial_l [jk,i] + [il,r]\Gamma^r_{jk} - [ik,r]\Gamma^r_{jl} + \end{aligned}$$

$$\partial_l [kj, i] - \partial_j [kl, i] + [ij, r] \Gamma^r_{kl} - [il, r] \Gamma^r_{kj} +$$
$$\partial_j [lk, i] - \partial_k [lj, i] + [ik, r] \Gamma^r_{lj} - [ij, r] \Gamma^r_{lk}$$
$$= \partial_k [jl, i] - \partial_l [jk, i] + [il, r] \Gamma^r_{jk} - [ik, r] \Gamma^r_{jl} +$$
$$\partial_l [jk, i] - \partial_j [kl, i] + [ij, r] \Gamma^r_{kl} - [il, r] \Gamma^r_{jk} +$$
$$\partial_j [kl, i] - \partial_k [jl, i] + [ik, r] \Gamma^r_{jl} - [ij, r] \Gamma^r_{kl}$$
$$= 0$$

where equality 1 is the block symmetry in the last term, equalities 2 and 3 are antisymmetry in the second term, equality 4 is the definition of the Riemann-Christoffel curvature tensor with required relabeling of indices, and equality 5 is the symmetry of the Christoffel symbols in their paired indices.

29. What is the pattern of the indices in the second Bianchi identity?
    Answer: The pattern is that the first two indices are fixed while the last three indices are cyclically permuted in the three terms.
30. Write the determinantal form of the Ricci curvature tensor of the first kind.
    Answer:
    $$R_{ij} = \begin{vmatrix} \partial_j & \partial_a \\ \Gamma^a_{ij} & \Gamma^a_{ia} \end{vmatrix} + \begin{vmatrix} \Gamma^a_{bj} & \Gamma^a_{ba} \\ \Gamma^b_{ij} & \Gamma^b_{ia} \end{vmatrix}$$
31. Starting from the determinantal form of the Ricci curvature tensor of the first kind, obtain the following form of the Ricci curvature tensor with justification of each step in your derivation: $R_{ij} = \partial_j \partial_i (\ln \sqrt{g}) + \Gamma^a_{bj} \Gamma^b_{ia} - \frac{1}{\sqrt{g}} \partial_a (\sqrt{g} \Gamma^a_{ij})$.
    Answer: Using the determinantal form of the Ricci curvature tensor of the first kind (which is given in the answer of the previous question) we have:
    $$\begin{aligned}
    R_{ij} &= \begin{vmatrix} \partial_j & \partial_a \\ \Gamma^a_{ij} & \Gamma^a_{ia} \end{vmatrix} + \begin{vmatrix} \Gamma^a_{bj} & \Gamma^a_{ba} \\ \Gamma^b_{ij} & \Gamma^b_{ia} \end{vmatrix} \\
    &= \partial_j \Gamma^a_{ia} - \partial_a \Gamma^a_{ij} + \Gamma^a_{bj} \Gamma^b_{ia} - \Gamma^a_{ba} \Gamma^b_{ij} \\
    &= \partial_j \partial_i (\ln \sqrt{g}) - \partial_a \Gamma^a_{ij} + \Gamma^a_{bj} \Gamma^b_{ia} - [\partial_b (\ln \sqrt{g})] \Gamma^b_{ij} \\
    &= \partial_j \partial_i (\ln \sqrt{g}) + \Gamma^a_{bj} \Gamma^b_{ia} - \partial_a \Gamma^a_{ij} - \Gamma^b_{ij} \partial_b (\ln \sqrt{g}) \\
    &= \partial_j \partial_i (\ln \sqrt{g}) + \Gamma^a_{bj} \Gamma^b_{ia} - \partial_a \Gamma^a_{ij} - \Gamma^a_{ij} \partial_a (\ln \sqrt{g}) \\
    &= \partial_j \partial_i (\ln \sqrt{g}) + \Gamma^a_{bj} \Gamma^b_{ia} - \partial_a \Gamma^a_{ij} - \Gamma^a_{ij} \frac{1}{\sqrt{g}} \partial_a \sqrt{g} \\
    &= \partial_j \partial_i (\ln \sqrt{g}) + \Gamma^a_{bj} \Gamma^b_{ia} - \frac{\sqrt{g}}{\sqrt{g}} \partial_a \Gamma^a_{ij} - \Gamma^a_{ij} \frac{1}{\sqrt{g}} \partial_a \sqrt{g} \\
    &= \partial_j \partial_i (\ln \sqrt{g}) + \Gamma^a_{bj} \Gamma^b_{ia} - \frac{1}{\sqrt{g}} (\sqrt{g} \partial_a \Gamma^a_{ij} + \Gamma^a_{ij} \partial_a \sqrt{g}) \\
    &= \partial_j \partial_i (\ln \sqrt{g}) + \Gamma^a_{bj} \Gamma^b_{ia} - \frac{1}{\sqrt{g}} \partial_a (\sqrt{g} \Gamma^a_{ij})
    \end{aligned}$$

    where line 2 is the expansion of the determinantal form, line 3 is the identity $\Gamma^j_{ij} = \partial_i (\ln \sqrt{g})$, line 4 is reordering of terms, line 5 is relabeling of dummy index in the last term, line 6 is the rule of differentiation of natural logarithm, line 7 is multiplying the third term with 1, line 8 is taking common factor from the last two terms, and line 9 is the product rule of differentiation.

32. Verify the symmetry of the Ricci tensor of the first kind in its two indices.
Answer: We have:

$$\begin{aligned}
R_{ij} &= \partial_j \partial_i (\ln \sqrt{g}) + \Gamma^a_{bj} \Gamma^b_{ia} - \frac{1}{\sqrt{g}} \partial_a \left(\sqrt{g} \Gamma^a_{ij} \right) \\
&= \partial_i \partial_j (\ln \sqrt{g}) + \Gamma^a_{bj} \Gamma^b_{ia} - \frac{1}{\sqrt{g}} \partial_a \left(\sqrt{g} \Gamma^a_{ij} \right) \\
&= \partial_i \partial_j (\ln \sqrt{g}) + \Gamma^a_{ib} \Gamma^b_{aj} - \frac{1}{\sqrt{g}} \partial_a \left(\sqrt{g} \Gamma^a_{ij} \right) \\
&= \partial_i \partial_j (\ln \sqrt{g}) + \Gamma^a_{bi} \Gamma^b_{ja} - \frac{1}{\sqrt{g}} \partial_a \left(\sqrt{g} \Gamma^a_{ji} \right) \\
&= R_{ji}
\end{aligned}$$

where line 1 is the result that we obtained in the previous question, line 2 is the commutativity of the partial differential operators in the first term, line 3 is reordering in the second term with relabeling of the dummy indices, line 4 is the symmetry of the Christoffel symbols in their lower indices, and line 5 is the result that we obtained in the previous question with relevant index labeling (i.e. line 4 is identical to line 1 with an exchange of i and j).

33. What is the number of distinct entries of the Ricci curvature tensor of the first kind?
Answer: Because the Ricci curvature tensor of the first kind is symmetric, then the number of its distinct[40] entries is:

$$N_{\text{RD}} = \frac{n(n+1)}{2}$$

34. How can we obtain the Ricci curvature scalar from the covariant Riemann-Christoffel curvature tensor? Write an orderly list of all the required steps to do this conversion.
Answer: We do the following:
(a) We obtain the mixed Riemann-Christoffel curvature tensor by raising the first index of the covariant Riemann-Christoffel curvature tensor, that is:

$$R^a{}_{ijk} = g^{ab} R_{bijk}$$

(b) We obtain the covariant Ricci tensor by contracting the contravariant index with the last covariant index of the mixed Riemann-Christoffel curvature tensor, that is:

$$R_{ij} = R^a{}_{ija}$$

---

[40] In fact, they are not necessarily distinct but they are independent, hence "distinct" in such contexts means "independent". This similarly applies to the entries of anti-symmetric tensors where distinct in such contexts means independent and hence an entry and its negative are not distinct (i.e. not independent) although they are distinct in a generic sense; moreover some independent entries may not be distinct in a generic sense.

(c) We obtain the mixed Ricci tensor by raising the first index of the covariant Ricci tensor, that is:
$$R^k{}_j = g^{ki} R_{ij}$$

(d) We obtain the Ricci curvature scalar $\mathcal{R}$ by contracting the indices of the mixed Ricci tensor, that is:
$$\mathcal{R} = \delta^j_k R^k{}_j = R^j{}_j$$

35. Make a list of all the main properties of the Ricci curvature tensor (rank, type, symmetry, etc.) and the Ricci curvature scalar.
    Answer:
    Some of the main properties of the Ricci curvature tensor are:
    (a) It is derived from the Riemann-Christoffel curvature tensor.
    (b) It is absolute rank-2 tensor.
    (c) It depends only on the metric tensor.
    (d) It is used to characterize the space and express its curvature.
    (e) It can be covariant (first kind) of type $(0, 2)$ or mixed (second kind) of type $(1, 1)$.
    (f) The covariant type is symmetric.
    Some of the main properties of the Ricci curvature scalar are:
    (A) It is derived from the Ricci curvature tensor (and ultimately from the Riemann-Christoffel curvature tensor).
    (B) It is absolute rank-0 tensor (i.e. scalar).
    (C) It depends only on the metric tensor.
    (D) It is used to characterize the space and express its curvature.

36. Outline the importance of the Ricci curvature tensor and the Ricci curvature scalar in characterizing the space.
    Answer: Because the Ricci curvature tensor and the Ricci curvature scalar are derived from the Riemann-Christoffel curvature tensor, they play similar roles to those of the Riemann-Christoffel curvature tensor in characterizing the space and depicting its curvature quantitatively and qualitatively. Hence, they have important uses and applications in several mathematical branches and physical theories like differential geometry and general relativity.

37. Write, in tensor notation, the mathematical expressions of the following tensors in Cartesian coordinates defining all the symbols involved: infinitesimal strain tensor, stress tensor, first and second displacement gradient tensors, Finger strain tensor, Cauchy strain tensor, velocity gradient tensor, rate of strain tensor and vorticity tensor.
    Answer:
    Infinitesimal strain tensor:
    $$\gamma_{ij} = \frac{\partial_i d_j + \partial_j d_i}{2}$$
    where γ_{ij} is the infinitesimal strain tensor and the indexed d is the displacement vector.
    Stress tensor:
    $$T_i = \sigma_{ij} n_j$$

where T_i is the traction vector, σ_{ij} is the stress tensor and n_j is the normal vector.
First and second displacement gradient tensors:

$$E_{ij} = \frac{\partial x_i}{\partial x'_j} \qquad \Delta_{ij} = \frac{\partial x'_i}{\partial x_j}$$

where E_{ij} is the first displacement gradient tensor, Δ_{ij} is the second displacement gradient tensor and the indexed x and x' are the Cartesian coordinates of an observed continuum particle at the present and past times respectively.
Finger strain tensor:

$$B_{ij} = \frac{\partial x_i}{\partial x'_k} \frac{\partial x_j}{\partial x'_k}$$

where B_{ij} is the Finger strain tensor and the other symbols are as defined before.
Cauchy strain tensor:

$$B_{ij}^{-1} = \frac{\partial x'_k}{\partial x_i} \frac{\partial x'_k}{\partial x_j}$$

where B_{ij}^{-1} is the Cauchy strain tensor and the other symbols are as defined before.
Velocity gradient tensor:

$$[\nabla \mathbf{v}]_{ij} = \partial_i v_j$$

where $\nabla \mathbf{v}$ is the velocity gradient tensor while $\mathbf{v}$ and v_j represent the velocity vector.
Rate of strain tensor:

$$S_{ij} = \frac{\partial_i v_j + \partial_j v_i}{2}$$

where S_{ij} is the rate of strain tensor and the indexed v is as defined before.
Vorticity tensor:

$$\bar{S}_{ij} = \frac{\partial_i v_j - \partial_j v_i}{2}$$

where $\bar{S}_{ij}$ is the vorticity tensor and the indexed v is as defined before.

38. Which of the tensors in question 37 are symmetric, anti-symmetric or neither?
    Answer:
    Infinitesimal strain tensor: symmetric.
    Stress tensor: not necessarily symmetric although it can be.
    First and second displacement gradient tensors: neither.
    Finger strain tensor: symmetric.
    Cauchy strain tensor: symmetric.
    Velocity gradient tensor: neither.
    Rate of strain tensor: symmetric.
    Vorticity tensor: anti-symmetric.

39. Which of the tensors in question 37 are inverses of each other?
    Answer:
    The first and second displacement gradient tensors are inverses of each other.
    Finger strain tensor and Cauchy strain tensor are inverses of each other.

7 TENSORS IN APPLICATION   151

40. Which of the tensors in question 37 are derived from other tensors in that list?
    Answer:
    Finger strain tensor is derived from the first displacement gradient tensor.
    Cauchy strain tensor is derived from the second displacement gradient tensor.
    The rate of strain tensor is derived from the infinitesimal strain tensor.
    Since the First and second displacement gradient tensors are inverses of each other, they may be considered as derived from each other.
    Since the Finger strain tensor and the Cauchy strain tensor are inverses of each other, they may be considered as derived from each other.
    Since the velocity gradient tensor is the sum of the rate of strain tensor and the vorticity tensor, it may be considered as derived from these tensors.
    Since the rate of strain tensor is the symmetric part of the velocity gradient tensor, it may be considered as derived from the velocity gradient tensor.
    Since the vorticity tensor is the anti-symmetric part of the velocity gradient tensor, it may be considered as derived from the velocity gradient tensor.
41. What is the relation between the first and second displacement gradient tensors?
    Answer: As indicated earlier, they are inverses of each other, that is:
    $$E_{ik}\Delta_{kj} = \delta_{ij}$$
    where δ_{ij} is the Kronecker delta and the other symbols are as defined earlier.
42. What is the relation between the velocity gradient tensor and the rate of strain tensor?
    Answer: As indicated earlier, the rate of strain tensor is the symmetric part of the velocity gradient tensor, that is:
    $$\mathbf{S} = \frac{\nabla\mathbf{v} + (\nabla\mathbf{v})^T}{2}$$
    where $\mathbf{S}$ is the rate of strain tensor and the other symbols are as defined earlier.
43. What is the relation between the velocity gradient tensor and the vorticity tensor?
    Answer: As indicated earlier, the vorticity tensor is the anti-symmetric part of the velocity gradient tensor, that is:
    $$\bar{\mathbf{S}} = \frac{\nabla\mathbf{v} - (\nabla\mathbf{v})^T}{2}$$
    where $\bar{\mathbf{S}}$ is the vorticity tensor and the other symbols are as defined earlier.
44. What is the relation between the rate of strain tensor and the infinitesimal strain tensor?
    Answer: The rate of strain tensor is the time derivative of the infinitesimal strain tensor, that is:
    $$\mathbf{S} = \frac{\partial \boldsymbol{\gamma}}{\partial t}$$
    where $\mathbf{S}$ is the rate of strain tensor, $\boldsymbol{\gamma}$ is the infinitesimal strain tensor and t is time.

45. What are the other names given to the following tensors: stress tensor, deformation gradient tensors, left Cauchy-Green deformation tensor, Cauchy strain tensor and rate of deformation tensor?
Answer:
Stress tensor: also known as Cauchy stress tensor.
Deformation gradient tensors: also known as displacement gradient tensors.
Left Cauchy-Green deformation tensor: also known as Finger strain tensor.
Cauchy strain tensor: also known as right Cauchy-Green deformation tensor.
Rate of deformation tensor: also known as rate of strain tensor.

46. What is the physical significance of the following tensors: infinitesimal strain tensor, stress tensor, first and second displacement gradient tensors, Finger strain tensor, velocity gradient tensor, rate of strain tensor and vorticity tensor?
Answer:
Infinitesimal strain tensor describes and quantifies the state of strain usually in continuum media such as viscous fluids.
Stress tensor describes and quantifies the state of stress in physical objects. It is also used for example in transforming a normal vector to a surface to a traction vector acting on that surface.
First and second displacement gradient tensors describe and quantify the displacement of a physical object (e.g. particle of continuum) in its historical development (i.e. relationship between its present and past position).
Finger strain tensor describes and quantifies the state of strain in physical objects (e.g. continuum media) in its historical development. This can be inferred from the fact that it is derived from the first displacement gradient tensor.
Velocity gradient tensor describes and quantifies the gradient of velocity (i.e. its rate of variation over space) in physical objects such as liquids and gases.
Rate of strain tensor describes and quantifies the local rate at which neighboring material elements of a deforming continuum move with respect to each other.
Vorticity tensor describes and quantifies the local rate of rotation of a deforming continuum medium.

Index

Absolute
    derivative, 3, 73, 83, 107, 108, 111
    differentiation, 73, 107, 110, 111
    permutation tensor, 5, 56, 77
    tensor, 39, 53, 58, 64, 66
Addition of tensors, 45, 46
Admissible transformation, 24, 25, 31
Affine tensor, 83, 85, 112
Algebraic
    addition, 45
    subtraction, 45
Anisotropic tensor, 40
Anti-
    symmetric, 41, 42, 44, 45, 54, 55, 140, 142, 150
    symmetric tensor, 44
    symmetry, 41, 43, 54
Area, 3, 79, 80, 137
Associate tensor, 72
Associative, 45
Associativity, 25

Basis
    tensor, 37, 46
    vector, 3, 4, 6, 10, 12, 17–19, 26, 29–32, 36, 51, 68, 83, 87, 97, 99, 100, 106–108, 112, 113
Bianchi identity, 146, 147
Block symmetric, 140–142
Bound index, 9

Cartesian, 3, 4, 21, 27–29, 37, 38, 47–49, 60, 67, 70, 71, 73, 79, 80, 82, 85, 93, 114–116, 130, 131, 136, 149
Cauchy
    -Green deformation tensor, 152
    strain tensor, 3, 149–152
    stress tensor, 152
Christoffel symbol, 3, 4, 72, 83, 85–87, 89, 91, 94, 96, 99, 100, 108, 120, 121
Circle, 18, 19, 27
Closure, 25
Comma notation, 3, 6
Commutative, 26, 45, 46, 48, 68, 102, 140
Composition of transformations, 25, 26
Conjugate tensor, 72
Conserved, 54
Continuity condition, 9, 24
Continuous, 9, 24, 104

Contraction, 47, 48, 50, 103, 104, 112, 113
Contravariant, 4
    basis vector, 3, 17, 19, 29–32, 51, 67, 68
    component, 29, 31, 120
    differentiation, 97
    index, 118, 119, 148
    metric tensor, 4, 32, 33, 70, 73
    permutation tensor, 5, 56
    tensor, 34, 35, 119
Coordinate
    curve, 17–19, 27–30, 32, 67, 69, 79, 82, 99
    surface, 17–19, 27–30, 32, 67, 69, 79, 80
    system, 4, 7, 10, 11, 15, 17–21, 26–30, 33, 34, 37–40, 52, 66–74, 77–83, 85, 86, 91, 92, 94, 97, 100, 108, 111, 114
Covariant
    basis vector, 3, 4, 17, 19, 29–31, 51, 67, 68, 87
    component, 29, 31
    derivative, 3, 83, 97, 99–103, 105, 107, 108, 118, 119, 139
    differentiation, 97, 100–102, 104, 110
    index, 113, 148
    metric tensor, 4, 31–33, 56, 67, 70, 72, 73, 81
    permutation tensor, 5
    Riemann-Christoffel curvature tensor, 139, 140, 142
    tensor, 34, 35, 75
Cross product, 77, 114, 130
Curl, 3, 39, 115, 116, 119, 120, 123, 125, 126, 129
Curve, 3, 4, 27, 76, 78, 107, 111
Curved space, 15–17, 72, 74, 139
Curvilinear coordinate system, 17–19, 29, 30, 86, 97
Cylindrical coordinate system, 3, 5, 18, 19, 21, 27, 28, 73, 79, 80, 82, 85, 92, 93, 95, 113, 124, 125

Determinant, 3, 4, 31, 32, 39, 40, 56, 58, 61, 70, 72, 78, 81, 130
Differentiable, 24, 99, 100, 107, 115, 118, 119, 124, 125, 127, 128, 137
Differential operator, 3, 35, 101, 113, 114, 140
Dimension of space, 3, 13, 16, 17, 20, 29, 40, 53–55
Direct product, 47
Displacement
    gradient tensor, 3, 5, 149–152
    vector, 3, 82
Distributive, 45

INDEX

Divergence, 3, 115, 118, 119, 122, 124, 126, 129
    theorem, 136
Dot product, 19, 32, 50, 68, 75, 115, 130
Dummy index, 6, 10, 12, 13
Dyad, 3, 37, 38

epsilon-delta identity, 65
Euclidean, 16, 17, 92

Finger strain tensor, 3, 149–152
Flat space, 15–17, 19, 72, 74, 139
Free index, 9–13, 35, 47, 138

General
    coordinate system, 4, 11, 26, 29, 36, 47, 67, 68, 70, 71, 77–81, 83, 105, 113, 116, 118, 119, 121
    tensor, 85, 108
Generalized Kronecker delta, 4, 62–64
Gradient, 3, 17, 29, 32, 67, 69, 106, 114, 117, 121, 124, 126–128
    vector, 32

Handedness, 27, 38
Homogeneous coordinate system, 19, 20, 74

Identity, 25
    tensor, 53
    transformation, 26
Imaginary, 19, 20, 74
Improper
    rotation, 40
    transformation, 27, 38, 39, 54
Index notation, 6
Indicial
    notation, 7, 8, 12, 102
    structure, 11, 12, 34
Infinitesimal, 3, 15, 27, 67, 81, 82
    strain tensor, 4, 149–152
Inner
    multiplication, 45, 48, 50, 72
    product, 3, 48, 50, 51, 72
Integral, 78
Intrinsic derivative, 109
Invariance, 7, 40, 44
Invariant, 7, 24, 38, 41, 44, 119, 134, 135
Inverse, 25
    Jacobian, 4
    of matrix, 130
    of metric, 33, 69, 70
    of tensor, 150
    transformation, 20, 21
Invertible transformation, 24

Isotropic tensor, 40, 53

Jacobian, 4, 20–22, 24–26, 72
    matrix, 4, 20, 21, 32

Kronecker delta, 4, 10, 53, 54, 59, 62, 63, 68, 70, 71, 102, 103, 110–112

Laplacian, 3, 114, 116, 121–125, 128
Length, 3, 4, 10, 15, 16, 67, 69, 74, 78
Levi-Civita tensor, 54
Linear algebra, 48
Lorentz transformations, 74
Lowering operator, 35

Manifold, 17
Matrix algebra, 48, 51
Metric tensor, 4, 15, 19, 26, 30–33, 53, 66–69, 72, 73, 78, 81, 85, 86, 92, 95, 96, 100, 101, 103, 110–112
Minkowski metric, 74
Mixed
    derivative, 104, 139
    Kronecker delta, 4
    metric tensor, 4, 68, 69
    Riemann-Christoffel curvature tensor, 139, 140, 142
    tensor, 4, 10, 14, 34, 68, 69, 117, 139, 142
Multiplication
    by scalar, 45
    of matrices, 47, 130
    of tensors, 45–48, 50, 72
Mutually
    orthogonal, 10, 19, 32, 68
    perpendicular, 17, 19

nabla operator, 3, 113–116, 122, 124
Negative orthogonal transformation, 27
Non-
    scalar tensor, 83, 97, 105, 106, 108
    singular, 66
Normal vector, 4, 150
Normalized vector, 3

Order
    of derivative, 104, 139
    of indices, 11–13, 35, 48, 71, 72, 94
    of multiplicands, 47, 101, 110
    of tensor, 10, 13
Orthogonal, 10, 17, 19, 32, 68
    coordinate system, 4, 17–20, 52, 67, 73, 78, 79, 81, 88, 91, 122, 124, 125, 127, 128
    transformation, 27, 38, 39

INDEX

Orthonormal
    basis set, 10, 31
        Cartesian, 28, 47, 60, 67, 70, 71, 73, 79, 80, 82, 93, 131
        coordinate system, 10, 28, 47, 60, 67, 70, 71, 73, 79, 80, 82, 93, 94, 131
        vectors, 3, 10, 30, 31, 36, 60
Orthonormalized vectors, 4, 36
Outer
    multiplication, 45
    product, 3, 50, 100, 101, 110

Parallelepiped, 31
Partial
    derivative, 3, 9, 24, 83, 87, 99, 100, 103, 104, 107
    differential operator, 35
    differentiation, 97, 100
Perimeter, 137
Permutation
    of tensor, 51
    tensor, 5, 54, 56, 77, 112
Perpendicular, 3, 17, 19, 29, 67
Physical
    basis vector, 51
    component, 52, 124
    dimension, 17
    representation, 3, 51, 52
Plane, 17, 18, 27
Polar coordinate system, 5
Position vector, 3, 4, 29, 78, 137
Positive orthogonal transformation, 27
Principle of invariance, 7
Product rule, 112
Proper transformation, 27, 54
Pseudo
    tensor, 38, 40
    vector, 38, 39

Quotient rule of tensor, 51

Raising operator, 35, 97, 121, 148, 149
Range of index, 6, 137
Rank
    -0 tensor, 8, 119
    -1 tensor, 8, 115
    -2 tensor, 8, 34, 37–40, 42, 43, 49, 53–55, 58, 66, 75, 98, 105, 115, 117, 119, 134, 135
    -3 tensor, 6, 37, 40, 54, 55
    -4 tensor, 37, 54, 55
    of tensor, 6, 8, 10, 40, 46, 54, 55, 99, 100, 112, 113

Rate of
    deformation tensor, 152
    strain tensor, 4, 149–152
Real, 20
Reciprocal, 21
Reciprocity relation, 37
Rectangular
    Cartesian, 37, 38
    coordinate system, 37, 38
Rectilinear coordinate system, 17–19, 108
Reflection, 27, 38
Relative
    permutation tensor, 5, 56, 77
    scalar, 105
    tensor, 4, 39, 56–58, 105
Replacement operator, 59, 110, 111
Ricci
    curvature scalar, 4, 148, 149
    curvature tensor, 4, 147–149
    theorem, 101, 102
Riemann-Christoffel curvature tensor, 4, 16, 17, 104, 105, 138, 140, 145–149
Riemannian
    geometry, 17
    space, 15
Right handed system, 30, 31
Rotation, 26, 27, 40

Scalar, 3, 6, 12, 13, 39–41, 45, 46, 58, 100, 105, 107, 112, 117–119
    field, 106, 117, 121, 123–126, 128
    invariant, 134, 135
    operator, 114
    triple product, 77, 81, 82, 130
Scale factor, 4, 28, 67, 74, 78–80, 82, 94, 124, 125
Schur theorem, 16
Semi-
    circle, 19, 27
    plane, 18, 19, 27
Semicolon notation, 3, 6
Shifting operator, 71, 102, 110, 111
Skew-symmetric, 42
Sphere, 19, 27
Spherical coordinate system, 3, 4, 18, 19, 22, 27, 28, 52, 73, 79, 80, 82, 85, 93, 95, 113, 124, 127, 129
Stokes theorem, 136
Straight line, 16–19, 27
Stress tensor, 5, 149, 150, 152
Subtraction of tensors, 45, 46
Sum rule, 112
Summation, 6, 10, 47

convention, 6
Surface, 3, 4, 17, 27, 137
Symbolic notation, 6–8, 102
Symmetric, 41, 42, 44, 53, 66, 68, 86, 91, 140, 142, 150
    tensor, 42, 44
Symmetry, 41, 43, 53, 66, 120, 121, 146, 148, 149

Tangent vector, 17, 29, 32, 67, 107
Tensor
    calculus, 52, 113, 130
    component, 12, 99, 113
    equality, 9, 11, 12, 40, 41, 46
    expression, 9, 11, 12, 38, 40
    field, 14, 107, 115
    multiplication, 46, 47
    notation, 35, 68, 78, 114–116, 123, 136, 149
    representation, 51
    term, 6, 9, 12, 13, 38, 40, 95, 99, 105, 147
    test, 51
Total
    derivative, 83, 107, 108, 111
    differentiation, 108, 109, 112
Totally
    anti-symmetric, 42, 55
    covariant Riemann-Christoffel curvature tensor, 139
    symmetric, 42
Trace, 4, 48, 130
Traction vector, 4
Transformation, 4, 8, 20, 21, 24–27, 31, 34, 38, 40, 41, 44, 51, 54, 70–72, 119
Translation, 27
True
    tensor, 38–40
    vector, 38, 39

Unit vector, 137
Unity tensor, 69

Vector, 3, 4, 6, 10, 12, 19, 29, 30, 34, 36–39, 41, 47, 51, 60, 68, 75, 77, 81, 97, 107, 114, 115, 117–119, 130, 137, 150
    calculus, 48, 113
    field, 114, 116, 119, 122, 124, 126, 129, 136, 137, 139
    identity, 136, 137
    notation, 136
    operator, 114
    triple product, 77, 130
Velocity
    gradient tensor, 3, 149–152
    vector, 4, 150
Volume, 3, 4, 31, 80, 81, 137
Vorticity tensor, 4, 5, 149–152

Weight of tensor, 4, 39, 46, 47, 56–58, 105

Zero tensor, 40, 41, 44, 85